Francesco Pietra

A Secret World

Natural Products of Marine Life

Springer Basel AG

Author's address:

Prof. Dr. Francesco Pietra
Instituto di Chimica
Università di Trento
38050 Povo-Trento
Italy

Illustrations by Gudrun Burgstaller

Library of Congress Cataloging-in-Publication Data

Pietra, Francesco, 1933 –
A secret world: natural products of marine life / Francesco Pietra.
   p. cm.
Includes bibliographical references

1. Marine pharmacology. 2. Biological products. I. Title.
RS160.7.P54   1990
574.92 – dc20

Deutsche Bibliothek Cataloging-in-Publication Data

Pietra, Francesco:
A secret world: natural products of marine life / Francesco Pietra. – Basel; Boston;
Berlin: Birkhäuser 1990

ISBN 978-3-7643-2346-2   ISBN 978-3-0348-7531-8 (eBook)
DOI 10.1007/978-3-0348-7531-8

© 1990 Springer Basel AG
Originally published by Birkhäuser Verlag in 1990.

The use of registered names, trademarks, etc. in this publication does not imply, even in
the absence of a specific statement, that such names are exempt from the relevant protective laws and regulations and therefore free for general use.

Printed from the author's camera-ready manuscript on acid-free paper

CONTENTS

CONTENTS

CONTENTS

Worm: a vague term. Annelid insecticides, acorn-worm tumor inhibitors, and plasticizers.

Adaptation and defense of barnacles, shrimps, and horseshoe crabs. Luminous seed shrimps and krill.

Cytotoxic substances of bryozoans. Repellency of brachiopods: a problem beyond the limits of present technology?

Pigments of crinoids and sea urchins. Fertilization, EPA, and growth factors in sea urchins: the impact on human affairs. Similarity in chemical armor of sea stars, sea cucumbers, and brittle stars. Growth and death of coral reefs: the *Acanthaster* phenomenon.

Sea squirts as specialized sinks for metals in unusual states. Pharmacology with aplousobranch sea squirts. Comparative biochemistry as a surrogate for fossil records.

Of fish and worms. Non-immune defense of fish. Below and above fishes in the evolutionary scale.

Eukaryotic vs. prokaryotic cells. Cell type and evolutionary marks. Photosynthesis in the sea. Carbon vs. silicon as a basis of life. The accessory pigments of photosynthesis: a phylogenetic mark for red seaweeds. Bacteriochlorophylls: the non-oxygenic photosynthesis. The nature of the photosynthetic process. Screening of natural products for biological action. Biological classification.

CONTENTS

Preface

Underwater exploration is a fascinating activity because of the richness of sea life, the beautiful colors, and the contrast with terrestrial life. It can be done by diving or snorkeling -today there is fine equipment for diving and taking underwater pictures-and, apart from being a source of pleasure, it forms the basis of modern ecological and behavioral marine studies.

Diving is an emotional activity. The silence of the submarine world, only interrupted by breathing, has more than once evoked the monsters of the fabulous tales of my youth. Maybe because, although much time has elapsed since then, I still have a youth's ability to evaluate everything for the fascination it conveys. However, no more than that can be gained from underwater watching. I am not saying that this is destined to be a futile exercise. Picasso watched sea urchins much in the same way while revealing their quintessence in his paintings at the Castle of Antibes. What I mean is that today marine life can also be viewed from the perspective of the natural products involved. This knowledge, subsequent to Picasso's time, adds much to the overall picture.

The importance of marine life, even in regard to man's short-term needs, can hardly be overestimated as the sea covers, without interruption, over 70% of the Earth's surface. The contribution of marine primary producers to the Earth's fixed carbon budget is comparable to that of terrestrial plants; moreover, the sea is a net oxygen producer owing to irreversible fixation of carbon dioxide, and nearly 80% of the Earth's animal life is in the oceans.

Thus, we are going to depend more and more on sea resources as the continents become more and more populated. But this is an old trend. Looking back, man has always been strongly dependent on the sea. Major populations have always lived along marine coasts and even recent major growths have occurred in such areas. It has been estimated that by the end of the century over 70% of the population of the United States will inhabit marine or Great Lakes coastal zones.

This book describes marine life from the viewpoint of the natural products specific to certain organisms. We are only marginally concerned here with the metabolites common to all forms of life, in the sea and on the Earth, about which many popular books have already been written. This is **not** a scholarly work; it is for the nonexpert. Its structure is such that it can be read in whole or in part, and the illustrations are only intended to convey a rough idea of the marine organisms discussed here. But for those who wish for more depth, I have provided appendices where the marine natural products which have a recognized ecological role, or which may prove useful to man, are represented at the molecular level.

This book is unique in its structure and scope. There is, in fact, very little information available at a popular level even on natural products of terrestrial organisms; a striking circumstance when one recalls the epoch-making discovery of penicillin, a mold product which has allowed us to live better and which accounts, together with other antibiotics of the same class, for billions of dollars in annual sales.

Who is the nonexpert this book is aimed at? He is one interested in Nature, in particular the diver, or someone wishing to compare marine with terrestrial life, or even someone involved in either the food, the drug, or the cosmetic industry. There is in fact an enormous market for marine products, not limited to fish, crustaceans and algae, but also comprising animal food, dung, algal and animal products, and this market is expected to increase considerably in the near future. If combined with the appendices, this book should also make useful supplementary reading for biology students, in particular those enrolled in marine biology courses.

FRANCESCO PIETRA

Pisa, spring 1990

INTRODUCTION

I find it hard to isolate the sciences from the humanities as the search for the truth in science depends so much on us and the methods we use. A recent harsh attack on the philosophy of science by Theocharis and Psimopoulos, denying its place in the domain of science, does not serve either the search for the truth, nor, in its biased view, the quest for more funding in scientific research. It is also a curious attack in its broadness, as the philosophy of science is so multifaceted today that all possible positions, from Voltaire's skepticism to pure pragmatism, through the positivism of Spencer, are amply included.

That the sciences are hardly separable from the humanities is testified by the history of man. From the beginning man was impressed by natural phenomena, first terrified, in a natural religion, by lightning and hurricanes. Then, starting with the ancient Greek philosophers, scientific curiosity was stimulated by natural phenomena. If this did not occur already much earlier with the ancient Egyptians whose mystery, not penetrated by the Greeks and the Romans, has not been clarified by such a lucky discovery as the Rosetta Stone. Post-Renaissance interest in classical antiquity, the unexplored world, and scientific archeology was followed, in boundless curiosity, by the investigation of natural history. Thus, although exceptions existed, such as the conservative Archbishop Ussher, who, by literally interpreting the Bible, calculated that the Earth was created in the year 4004 B.C., people demanded for new discoveries at that time and explorations received support. This has continued and in the last century support was given not only to archeological expeditions (which have thrown so much light on the history of man, such as in the planned search, and discovery, of Homer's Troy by the inspired German Heinrich Schliemann) but also to scientific marine expeditions. Important cruises such as that of the English Challenger, guided by Sir John Murray from 1872 to 1876 for over 69,000 miles through the seven seas, or the Dutch Siboga, have revealed unknown worlds and drawings of living organisms at accurate scale in monumental reports from such expeditions are still much used today.

1

INTRODUCTION

Even in our time scientific cruises are essential to marine research. In France there are the cruises with the Marion Dufresne vessel in the sub-Antarctic area or the Seamount cruises in the Atlantic. There have also been scientific cruises specifically devoted to bioactive natural products, such as the American Alpha Helix expeditions in the Caribbean, or Soviet expeditions in the North Pacific. In these cases large vessels equipped for biological chemical research were used. Unfortunately, limited funds forced the Americans to interrupt such cruises and sell the Alpha Helix vessel.

Excepting the direct involvement of the private company Roche in Dee Why, on the Great Barrier, and the interest of Japanese companies, marine organisms have recently been collected for the study of natural products by scientist who are also divers. The old organization of marine scientific stations proved to be a fundamental support and most marine organisms have been collected in the areas of such institutions, as with taxonomic studies in the last century.

"How do you know what to collect?" is a question the marine natural product chemist is often asked. There is no simple answer: it is the result of an awareness, gained from previous studies, concerning the distribution of natural products in marine taxa, and of field observations, such as the presence of associated organisms or the lack of predation.

Private companies, not only from the USA, France, and Japan, but also Spain, are beginning to show interest in bioactive marine natural products. The Japanese seem most involved, stimulated by Japan's Ministry of International Trade and Industry in a ten-year project, which intends to establish new industries for extracting natural fine chemicals from the sea. Since the new companies operate on a large budget (in particular with their own well equipped vessels for long cruises and submersibles which have the ability to collect living organisms at a depth of 300-400 meters) the area of marine natural products will expand tremendously. In contrast, the university, equipped only for SCUBA diving at small depths, is expected to stay behind.

2

GENERAL INTRODUCTION

This book is faced with the problem that the term "marine living organisms" stands for a huge number of species. Therefore, although we are going to examine a small fraction of them, as nothing is known about the natural products of most species, we need a classification system. Non-animals are divided into three chapters; two chapters deal with primary producers, i.e. with the photosynthetic organisms which are the basis of life, beginning with microalgae and then seaweeds and higher plants, following the current taxonomic distribution. Chapters 4 and 5 are concerned with decomposers, bacteria and fungi, where we have to abandon their taxonomic classification as it is too intricate and would lead to dispersing the little we know about marine natural products from these taxa.

The chapters devoted to animals follow traditional taxonomic classification, starting with primitive animals such as sponges and ending with fish and reptiles.

The three appendices are for those who want to go beyond the popular level of chapters 1-14. Appendix A deals with biological fundamentals, such as cellular structures, photosynthesis and the structure of chlorophylls, and taxonomy. For the marine natural products mentioned in previous chapters, structural formulae are displayed in appendix B or a reference is made to closely related structural formulae[1]. Appendix C is an attempt to find a link between the classical taxonomy and the phylogenetic origin of marine organisms on the one hand and their secondary metabolites on the other. Reading these appendices will afford a much broader view of the subject, though they require some knowledge of biology and structural chemistry.

A glossary of terms frequently employed, which appears next, should be used in conjunction with the text as an aid for the more technical parts.

The book ends with a detailed index which groups together information which

---

[1] It is customary to assign trivial names to natural products as the scientific nomenclature is so cumbersome that it is only used for retrieval purposes. Where trivial names have not been assigned to natural products by the original author, I have taken the liberty to give them new names which appear within single quotation marks.

3

INTRODUCTION

is sparsely covered in the book, such as chemical compounds belonging to the same class but produced by unrelated living organisms.

1 An introduction to marine natural products

PRIMARY AND SECONDARY METABOLISM, TWO BADLY CHOSEN TERMS

If examined at the molecular level, the basic processes of life occur in much the same way with all living organisms. This is particularly apparent if we consider the animals and plants on the Earth's crust, including freshwater. All higher animals have analogous respiratory pigments and all higher plants use the same chlorophylls in photosynthesis. This leads to the same sugars and amino acids from all plants and the same proteins from all animals being formed. This is called primary metabolism and primary metabolites are the compounds that are involved in these processes.

However, it is also true that some plants can be safely eaten while other ones are poisonous and that the poison differs from species to species. Such poisonous compounds stem from processes that start with primary metabolites. Because of this, and as a consequence of the unfortunate choice of the term primary metabolism, these processes and chemical compounds have been called secondary metabolism and secondary metabolites.

Despite their name, secondary metabolites are by no means of secondary importance to the life of the organism. What is true is that secondary metabolites are in general restricted to certain organisms where they fulfill special roles, thus contributing to the uniqueness of these organisms as well as being essential to their survival.

To avoid difficulties the alternative term "natural products" was invented for the secondary metabolites. Secondary metabolites are not the only ones to have a natural origin, however: primary metabolites have a natural origin as well.

In lower forms of life different groups of bacteria and fungi vary morphologically, biochemically, and physiologically. This results in the production of a wealth of unusual natural products, many of which are used in medicine to alleviate our ailments.

Besides considering how man has used marine natural products to his advantage, we also have to examine in what way natural products are involved in

5

regulating marine life and thus contribute to the sharp differentiation between the marine and the terrestrial ecosystem. These ecological aspects are much less understood than nutritional and pharmacological ones and there is a good reason why: it is easier to obtain a grant for research based on our immediate needs than on marine ecology, although a thorough understanding of the latter could help us to live better in the long run.

MARINE VS. TERRESTRIAL NATURAL PRODUCTS

When comparing terrestrial with marine organisms we are struck by their considerable differences. Terrestrial higher plants include a huge number of species from which a wide variety of secondary metabolites can be extracted. Some of these plants are used in medicine or, regrettably, are the basis of drug abuse. In contrast, there are only a few species of higher plants in the sea and, moreover, they are a very scarce source of secondary metabolites. The situation is reversed when animals are examined; unusual natural products are much more commonly found in marine than in terrestrial animals. These are the facts, though rationalization is not easy; comparing terrestrial with marine plants and animals means to compare largely different taxonomic groups.

Many natural products derived from marine organisms display strong biological activities. Whether this will lead to new drugs which will alleviate our health problems, as with the natural products of terrestrial plants and microorganisms, is not yet clear. After the initial optimism of two decades ago, and bitter disillusionment a decade later, there is currently a more cautious, though concrete, approach toward bioactive marine natural products. Some pharmaceutical companies abandoned the site early, whereas other ones recently took up the challenge, and are discovering marine natural products with powerful biological activities. The future will do justice to those following the right way.

INTRODUCTION TO MARINE NATURAL PRODUCTS

What was immediately clear at the onset of these studies, and still remains true, is that many marine natural products are largely responsible for the appearance and regulation of marine life. For instance, some marine products induce spawning, or protect from other aggressive organisms, while the beautiful colors of marine organisms are to a large extent due to natural products. There are also many very hypothetical conclusions in this area, however. Most tests for biological activity have been carried out under artificial conditions that can hardly reproduce the natural conditions of the marine environment. Such investigations, in response to the scientist's urge to publish, often result in the search for shortcuts that obscure our attempts understand marine life even further.

SEAWATER IS SALTWATER

A specific condition under which marine species grow needs some comment. Seawater (the solvent) contains about thirty to thirty-five grams of dissolved salts per liter, mainly the common salt (the solute), while freshwater has a total salt content of only about one gram per liter. This results in an impressive phenomenon when marine water and freshwater are separated by a semipermeable membrane, as most animal membranes, such as the bladder of mammals, are. Such membranes inhibit the flow of most molecules, allowing the flow of water molecules to go in the direction of diluting the more concentrated solution (marine water, in this case) until the two solutions have reached the same concentration. This process is called osmosis and the pressure that builds up across the membrane increases with the concentration of the solute, i.e. with the number of solute particles per unitary volume. As a result, the osmotic pressure can be very high for solutes of low molecular weight, where a huge number of solute particles per unitary volume is present. Thus, with a semipermeable membrane that separates seawater from freshwater an osmotic pressure of about 25 atmospheres can build up. In contrast, with molecules of high molecular weight, like proteins, for the same weight of solute as in the case of the salt, the number of solute molecules per unitary volume is some thousand times smaller, and the osmotic pressure correspondingly is only a thousandth of that pertaining to seawater[2].

This phenomenon occurs at any time that two solutions at different concentrations of solute come into contact through a semipermeable membrane. For instance, if a human-blood cell, the walls of which are semipermeable and which contains about nine grams of common salt per liter, is placed in seawater it will collapse whereas in freshwater it will burst.

Consequently, the body fluids of aquatic organisms must have the same osmotic pressure of the water medium in which they live; such pressure is high in the sea and low in freshwater. A marine organism placed in freshwater will die, and the same will occur to a freshwater organism placed in seawater. A few animals can adjust their internal osmotic pressure quickly in response to external differences in salinity; they are called euryhaline animals.

To appreciate how the salinity of seawater has influenced the development of life, we have to forget about biblical chronologies by the Archbishop Ussher, and join the archaeologists P. C. Schmerling in Liège and Father MacEnery in Kent. Mid last century they proved, by finding relics of humans mixed with those of extinct animals, that man had lived during the same time as the mammoth. Darwin had a broader approach to the history of man. His book "Origin of species by means of natural selection" (published in 1859) is a landmark. If we accept Darwin's theory that species have evolved and that at the beginning life was in the sea, we have to admit that evolution took place under conditions of high internal osmotic pressure.

---

[2]The same is true for the freezing point of solutions, which is decreased by dissolved compounds, the more so the higher their concentration is. For the same weight, substances of low molecular weight exert a higher depression of the freezing point than large polymeric molecules, such as proteins. However, proteins have the ability of decreasing the freezing point of aqueous solutions by an independent mechanism, i.e. by preventing ice crystal growth down to -2.2 °C. This is important for the adaptation of marine organisms to the extreme climatic conditions of Arctic and Antarctic areas.

INTRODUCTION TO MARINE NATURAL PRODUCTS

There are other consequences related to the high salinity of seawater. Apart from chloride ion (which is one of the component particles of common salt, the other one being sodium ion), seawater also contains relatively large amounts of bromide and iodide ions (about 0.065 and 0.0005 grams per liter, respectively), two elements which are rare on the Earth's surface. Certain marine organisms have taken advantage of this and evolved unique enzymatic systems, called bromoperoxidases, that are capable of introducing bromine and iodine into natural products. A case in point is thyrsiferol, a brominated terpene of squalene origin which is produced by the red seaweed *Laurencia thyrsifera* (Hook) of New Zealand and a variety of

Laurencia obtusa of Teuri Island, Japan. On land or in freshwater only some microorganisms have a similar capacity which, moreover, is limited to chlorine.

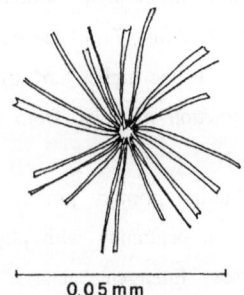

0.05 mm

Acantholithium dicopum

Seawater also contains calcium ions and the chemically very similar strontium ions. Whereas calcium ions are utilized by most living organisms, planktonic acantharians, such as *Acantholithium dicopum*, where strontium replaces calcium as skeletal component, are the sole species that utilize strontium ions. These organisms are very tiny but so abundant that they affect the global strontium level of the sea.

MARINE PRIMARY PRODUCERS

Macroscopic seaweeds and unicellular or colonial phytoplankton, collectively called algae, and seagrasses, which are flower plants, are the primary producers in the sea. With the effect of solar light, they are involved in the fixation of carbon dioxide (a gas present in the air, dissolved in the waters, and used in fizzy drinks) resulting in evolution of oxygen. This process is called photosynthesis.

Strictly speaking, the distinction between algae and vascular plants is very weak. Though the cell walls of seaweeds lack lignins (the polymers that make the higher plants woody), a vascular system similar to that of the higher plants is apparent in many algae. Moreover, the algae represent a polyphyletic group that is difficult to define. From the point of view of the secondary metabolism, green algae are closer to the higher terrestrial plants than to either brown or red algae. However, the distinction of algae from higher plants, having practical utility, is maintained here.

Marine primary producers are examined here according to their ecological distribution beginning with phytoplankton, via the seaweeds, and ending with the seagrasses. Emergent plants, such as *Heritiera littoralis* in salt marshes and mangrove swamps, can not be considered to be true marine plants but represent a borderline case, quite relevant to sea ecology, which warrants consideration here. Within each of these groups we adopt the traditional biological classification.

The sandy bottom of shallow-water marine areas is dominated by seagrasses, such as *Posidonia* in the Mediterranean, whereas rocky areas are dominated by seaweeds. Such vegetation is familiar not only to snorkelers and SCUBA divers, but also to people living on the coast. In temperate high-tide areas, such as Brittany, at the end of the growing season emergent rocks at low tide are covered by a dense mantle of seaweeds, which is a characteristic of the landscape. In contrast, phytoplankton, being small, escapes our attention except under special circumstances which are examined below.

10

2 Phytoplankton

Phytoplankton consists of microscopic algae, unicellular or colonial. The major populations belong to brown algae of the classes diatoms (which are encapsulated in a siliceous box and lack locomotion) and dinoflagellates (which are often reinforced by plates of cellulose and possess two flagella for locomotion). However, photosynthetic bacteria (cyanobacteria), with the behavior of algae, may sizably contribute to the phytoplankton population in tropical waters of low nutrient value, and brackish waters. The same is true for microscopic green algae (green flagellates) in coastal waters of temperate regions. Benthic phytoplankton, except in shallow waters where it occurs in a mixture of species, consists mainly of diatoms. A small group of marine phytoplankton is represented by the Chryptophyceae, which are more numerous in freshwater.

Let us now reflect on definitions. The term phytoplankton seems to imply photosynthetic organisms having planktonic life. While this is true in most cases, it does not apply to other notable cases. For example, the genus *Gambierdiscus*, which is central to our subsequent discussion, is benthic and some dinoflagellates, like the zooxanthellae, live in the body of marine invertebrates, such as sponges and corals. Parasitic algae and endolithic red (such as *Conchocelis*) and green algae (such as *Ostreobium*) also exist. Of course, all these are examples of non-planktonic algae. Finally, many cyanobacteria grow on shore rocks rising out of the water. Well, do not worry about terms and definitions; classifying such a complex matter as life must take some compromise in stride, such as considering all flagellates to be phytoplankton, in order to be of practical value.

In any event, phytoplankton are by far the major primary producers in the sea; they are responsible for about 95% of the primary productivity and net production makes up nearly half the gross production. However, production is uneven as phytoplankton reproduce by cell division or sexually, at rates that are seasonally dependent, with peaks in summer in Arctic and Antarctic areas and in winter in tropical waters.

MARINE PRIMARY PRODUCERS

Morphological adaptations have a bearing on the amount of sinking of planktonic algae; in warm, and thus low-density, waters, dinoflagellates have large projections which help to diminish the sinking rate. In contrast, in cold waters (e.g. in winter) algae adapt to the increased density of the water by reducing the size of their projections. Reduction in the size of phytoplankton also results in an increase of the surface area in relation to the volume, thus facilitating the uptake of solar rays and nutrients.

In addition to the use of the cysts of benthic dinoflagellates to date sediments for the oil industry, much importance is now given to the ability of phytoplanktonic cells of taking up toxic foreign substances, such as mercury compounds and pesticides; this occurs at rates that surpass dilution in the cell as a result of growth and division. This is detrimental to photosynthetic efficiency, and therefore to the growth of phytoplankton. Thus, though other toxic substances seem to be better tolerated by phytoplankton, discharging toxic compounds into the sea may have catastrophic consequences, owing to the fact that the food chain starts from phytoplankton. These problems should not be overlooked when planning to use phytoplankton for the detoxification of waters.

One has known for a long time that the growth of phytoplankton in the oceans is limited by the accessibility of reduced nitrogen rather than phosphorus[3]. Recently it has been suggested that also the availability of dissolved iron in the waters is a limiting factor and that this might have had an important effect on the history of the world. Thus, ice ages were periods during which there was an increased shift of dust from the land to the sea, and as dust is known to contain iron compounds, there might have been an overgrowth of phytoplankton during the

---

[3]Of certain North Atlantic and North Pacific areas which are downwind of heavily urbanized and industrialized zones the contrary is true: phosphorus is growth limiting. This seems to be the result of a large supply of reduced nitrogen in the form of nitrogen pollutants.

ice ages that subtracted carbon dioxide from the waters and air for the photosynthetic process. This is one suggestion for a lower level of carbon dioxide in the latest ice age than in the pre-industrial era. A diminished carbon dioxide level in the air allows more heat to escape from the atmosphere and therefore marine phytoplankton, resulting from abundant iron in the waters, might have contributed to the cooling of the world in ice ages. However, this requires that carbon dioxide is irreversibly subtracted from the waters and air, such as in the formation of oily sediments. Otherwise, carbon dioxide would be given back to the waters and air in the process of decomposition of organic matter.

FLAGELLATES AND OILY SEDIMENTS

The sterol composition of green flagellates is known to us. We know also that the growth rate of the diatom *Nitzschia frustulum* in culture is decreased by the presence of the green flagellate *Chlorella vulgaris*; the release of inhibitory metabolites by the green flagellate, to help its survival, has been suggested but never proved. But perhaps the largest impact of microscopic green algae on the environment of the sea has been their contribution to the formation of oil-rich deposits during the Ordovician period. Freshwater species, such as *Botryococcus brownii*, have maintained this ability of producing large amounts of terpenoidal hydrocarbons.

It is also believed that brown flagellates of the class Prymnesiophyceae may have contributed to the formation of organic Quaternary marine sediments. These sediments and contemporary flagellates of this class, like *Emiliania huxleyi*, contain long chain oxidized hydrocarbons.

WHY RED TIDES ARE RED

Phytoplankton is small and thus hard to see except under particular conditions, such as when it grows by division at high rate resulting in "red tides". This is a

13

phenomenon already noted in the Old Testament "And Moses and Aaron did so, as the LORD commanded; and he lifted up the rod, and smote the waters that were in the river, in the sight of Pharaoh, and in the sight of his servants; and all the waters that were in the river were turned to blood. And the fish that was in the river died; and the river stank, and the Egyptians could not drink of the water of the river; and there was blood throughout all the land of Egypt" (Exodus, 7, 20-21).

The expression "red tide" stems from the color of these phytoplankton which is due to pigments called carotenoids; these can be either free or bound to proteins as carotenoproteins. The main carotenoid of terrestrial plants is carotene, which is widely used as a food pigment and serves as an accessory pigment to gather light for photosynthesis in plants. Other carotenoids have the function of dissipating excessive solar radiation which would otherwise kill the photosynthetic organism. Phytoplankton also contains unique carotenoids, such as peridinin and fucoxanthin, which are the main pigments in dinoflagellates and diatoms, respectively, and which are structurally more complex than carotene. All carotenoids are labile compounds which undergo rapid modifications in coastal waters by reacting with molecular oxygen in a process which is induced by solar radiation.

You may have noticed that, after the general introduction to halogenated marine natural products in Chapter 1, this is the second time that specific differences between marine and terrestrial natural products are apparent. Further reading will reveal that such differences are the rule rather than the exception. There is some hint, however, that steroids could be a case apart: many of the unusual steroids of sponges, which were thought up to now unique to this phylum, seem to be present in terrestrial plants as well. However, as steroids are fundamental to life it is no wonder that they are widely-spread.

The carotenoids offer the opportunity to clarify other problems concerning natural products. I have mentioned above that the red color of phytoplankton may be due to either carotenoproteins or to free carotenoids. Proteins are biopolymers which function as enzymes, i.e. as catalysts of metabolic processes, and structural elements.

Most of the beautiful colors of marine invertebrates are due to carotenoproteins and it is still uncertain as to whether carotenoids occur as free molecules in living organisms on a general basis. The reason why greater attention has been paid to carotenoids than to carotenoproteins is that extractions with organic solvents from living organisms lead to carotenoids and not to carotenoproteins. Extraction processes are crude and result in breaking the labile bonds between the carotenoid and the protein; this methodology is therefore unsuitable in distinguishing between free carotenoids and carotenoproteins.

Whatever the form in which they occur and role, are the carotenoids secondary or primary metabolites? They are traditionally classified as secondary metabolites though their wide distribution in vegetables and generalized role in vital processes, which are typical attributes of primary metabolites, are widely recognized. Again it is good that I am not worried by definitions.

In any event, the actual form in which secondary metabolites occur in living organisms represents a general problem. In many cases what is finally isolated is not what was originally present in the living organism, which makes the study of the functions of natural products particularly difficult. A special case is that of secondary metabolites which are stored, in a protected form, in specific glands of living organisms and are set free in harmful form for defensive or offensive purposes, following enzymatic reactions. This occurs in terrestrial arachnids called opiliones, for example, some of which live in our houses and which, like *Opilio canestrinii*, resemble spiders with long legs and are called daddy-long-legs.

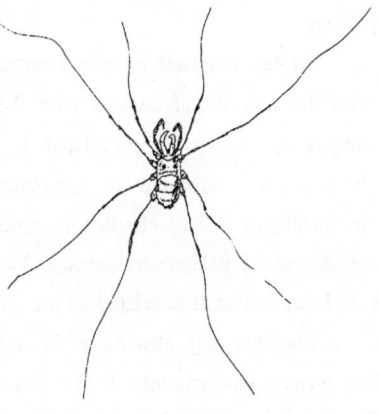

Opilio canestrinii

MARINE PRIMARY PRODUCERS

THE MILKY SEA AND OTHER STORIES OF BIOLUMINESCENCE

So much about carotenoids and carotenoproteins. Another noticeable phenomenon is the bioluminescence of phyto- and zooplankton, which makes the sea look alive on dark nights. Though emission of radiation is common to all biological systems (most of us are familiar with fireflies or luminous mushrooms) bioluminescence is a typical characteristic of marine organisms, which is diffused in nearly all taxa.

The notion as to why there are so many bioluminescent marine organisms is that luminescence serves to attract prey, or as a signal within the species, or even for protective purposes, camouflaging the outline of the organism during the day. However, the reason why bioluminescence first arose is probably different. Most bioluminescent processes consume oxygen and thus it could be that they initially served to eliminate traces of oxygen to make the atmosphere perfectly anoxic, as was required by the early living organisms.

In any event, bioluminescent processes are finely tuned. The yield of light per energy consumed is much higher than we have been able to attain with any artificial process.

Bioluminescence on ocean surfaces is mostly due to dinoflagellates and results from the reaction of oxygen with the luciferin of these algae (the bioluminescent compound), which is stimulated by a luciferase (the enzymatic system). The phenomenon is now clearly understood at the molecular level: the luciferin of the dinoflagellates ('dinoluciferin') is structurally related to the luciferin of krill, which are shrimplike pelagic crustaceans. The structure of krill and dinoflagellate luciferin is unique, in that it is related to the tetrapyrrole unit of heme.

Plankton may also make the seawater so turbid that visibility is reduced to a few meters. One example is the Côte Vermeille of Mediterranean France, near the French border, where benthic species, in the search for light, live closer to the

surface than in clear waters with
little plankton, such as the arid
coasts of the Red Sea or the
Aegean part of the Mediterranean.
As a result of and in order to have
access to a wealth of marine
invertebrates and algae in very
shallow waters, a marine scientific
station was founded in the last
century in Banyuls-Sur-Mer, on the
Côte Vermeille.

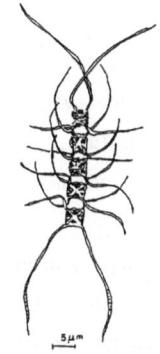

Chaetoceros

THE DISTRIBUTION OF DIATOMS

Diatoms dominate in Arctic and Antarctic waters and in temperate areas, where they
may form blooms except in summer when they
are replaced by flagellates. In cold areas
phytoplankton samples may reveal up to 400
different diatoms. *Chaetoceros*, *Navicula*, and
Nitzschia are common genera which can be
found both in the Mediterranean and along the
north and middle Atlantic coasts of the United
States.

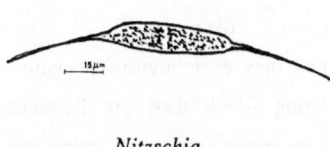

Nitzschia

Much of what has been said about phytoplankton in general applies to
diatoms. In particular, diatoms are able to adapt to the increased density and
viscosity of seawater in winter by thickening their cell walls and shortening cell
projections in order to increase the sinking rate. Their cell division rate and
photosynthetic activity increase with the decreasing temperature of the sea.

DIATOMS AND ATHEROSCLEROSIS

EPA, a fatty acid of widespread occurrence (found in fish oil for example) is structurally similar to terrestrial vegetable oils and has valuable antithrombic and antiatherosclerotic properties. EPA, which is also produced by diatoms, such as *Phaeodactylum tricornutum*, originates biosynthetically from acetic acid and is therefore an acetogenin.

A biotechnological company in the Hawaiian island of Kona has started to produce EPA and carotene from cultures of microalgae favored by climatic conditions. At the time that this book was written, the company was expecting to capture a sizable fraction of the world market with these compounds, perhaps 2-3% of the carotene market, which amounts to 50 million dollars.

DIATOMS, PENGUINS, AND PHARMACEUTICAL COMPANIES

The common diatom *Skeletonema costatum* contains an unstable compound, ß-dimethylpropiotethin, which breaks up to form a small sulfurated compound, dimethyl sulfide, together with an antibacterial organic acid, acrylic acid. Curiously, this is the acid used by the textile industry producing acrylic fibers.

Acrylic acid also occurs with other diatoms and may even involve penguins. These animals have seasonal digestive problems during which their gut becomes bacterially sterile. This results from ingesting planktonic crustaceans which retain the acrylic acid precursor of *Phaeocystis costatum* and other diatoms on which they feed. Acrylic acid is thus released into the gut of penguins where it inhibits the bacteria which are needed to digest food thoroughly.

The *in vitro* antibacterial activity of a derivative of EPA, EPA phytol ester, is also known. This compound is produced in cultures of the diatom *Navicula delognei* var. *elliptica* collected along the New Brunswick Canadian coasts. A similar activity of an undefined polymeric sugar isolated from cultures of the Mediterranean diatom

Chaetoceros lauderi has also been reported. These are examples of bioactive compounds that raise less interest today than in the past. The reason is that not only are there good antibiotics already on the market, but *in vitro* antibacterial activities, as in the above cases, may be far from clinical use. The antibacterial compound may simply have local rather than systemic activity and, even worse, may prove toxic to man, or the compound may be unstable, which makes commercial stocking difficult.

Faced with so many problems and the long and expensive procedures of toxicity tests, which are necessary before pharmaceutical products can be approved by state health organizations, many pharmaceutical companies have lost interest in developing new antibacterial products and prefer to continue the production of well established antibiotics. Prospects for the future, when pathogenic bacteria will have acquired resistance to current antibiotics, are cloudy. We are already finding it difficult to control certain bacterial infections, particularly certain mycobacteria. Thus, while *Mycobacterium tuberculosis* and *Mycobacterium leprae* are always pathogenic for man (they are responsible for tuberculosis and leprosy) other species, such as *Mycobacteriun avium*, are opportunistic and normally non-pathogenic. However, with AIDS patients, where the immune system is weakened, antibiotic-resistant mutants of such bacteria may become pathogenic and bring forth pulmonary tuberculosis which can not be controlled.

Though it is clear that there are areas where new antibiotics are needed, medical interest in natural products is now focused on the treatment of diseases that are not yet well understood. Compounds that have antiviral, vascular and cytotoxic activities are deemed worth considering by most pharmaceutical companies and there is an increasing interest in agents active against parasites and insects which are present in animals, or that act as plant growth stimulators, or as enzyme inhibitors or activators.

ARE DINOFLAGELLATES CAPABLE OF LEARNING?

It takes time to recover from a west to east flight when the night rapidly merges with the day, no matter how comfortable the flight and the warm croissants for breakfast, like in a parisian café.

The time it takes to recover from the flight is the time it takes our internal clock to readjust to local conditions. This is not our exclusive problem or the problem of organ-based animals only. It occurs in plants and even in unicellular organisms like the dinoflagellate *Gonyaulax polyhedra*. For reasons that are poorly understood, *G. polyhedra*

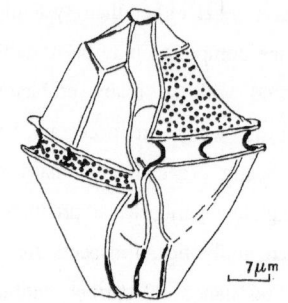

Gonyaulax polyhedra

emits light at regular intervals during the night, with precise regularity over 24 hours. The extraordinary fact is that when the dinoflagellate is kept in continuous darkness this phenomenon continues unaltered; *G. polyhedra* has learned the art of being a nocturnal light emitter too well.

These endogenous rhythms with periods of about 24 hours are called circadian rhythms from the Latin *circa* (about) and *dies* (day) and have nothing to do with exogenous diurnal rhythms, which should actually be called daily rhythms.

Cellular extracts taken at regular intervals from *G. polyhedra* revealed that the concentrations of luciferin, the light-active compound, and luciferase, the light-inducing enzyme, are higher during the night than during the day. Precise control of the enzyme activity and, in consequence, of light emission is probably exerted by a regulatory metabolite, which is released in tiny amounts from cell organelles of the dinoflagellate for a certain period and then switched off for the next period when it is no more needed. Substances with such properties, called hormones, are at the basis of metabolic control in all organisms, including man.

POISONOUS DINOFLAGELLATES: FROM PARALYTIC SHELLFISH
POISONING TO THE BREVETOXINS

Dinoflagellates occur as single cells or as colonies and dominate phytoplankton in warm waters; in the Mediterranean they are abundant in summer. In other temperate areas, during the warmer season, they are able to replace diatoms.

Some dinoflagellates have photosynthetic ability, but can also utilize solid food (e.g. bacteria) like animals, so that there was a long dispute as to whether the dinoflagellates are of more interest to the botanists or the zoologists. In any event, certain dinoflagellates are largely responsible for bioluminescence in the sea on dark nights; this occurs in flashes, a fact that is reflected in the species' name *Noctiluca scintillans*.

Dinoflagellates divide at a high rate during certain periods and either form blooms (red tides) or settle on benthic organisms. A few species are toxic and the toxins can be passed on via the food chain to mollusks and fish and may kill fish and birds en masse. In turn, intoxicated mollusks and fish may intoxicate man. This is a subtle danger seeing as the same fish or mollusks are safely edible during other periods. Although, outside the tropics, car traffic is far more dangerous than toxic fish, sporadic cases of deadly intoxications have occurred far away from the tropics since frozen food was commercialized. Even in Venice there was a case of tourists dying after they had eaten tropical fish containing dinoflagellate toxins in a restaurant.

Outbreaks of dinoflagellates are of considerable local concern and have therefore received much attention. The responsible genera can be divided into three groups. The first group comprises dinoflagellates of the genera *Gonyaulax* and *Protogonyaulax* (order Peridiniales) whose toxins (saxitoxin and gonyautoxins, also named paralytic shellfish poisons) accumulate in the hepatopancreas of mollusks and fish. Outbreaks of *Gonyaulax tamarensis* and luminescent and non-luminescent strains of *G. excavata* often occur along the Atlantic coasts of Great Britain, Canada and the United States. In the northeastern Pacific and Japan, blooms were due to *Protogonyaulax* species whereas *Gonyaulax catenella* was responsible for blooms off the Pacific coast in North America.

Extractions from either intoxicated mussels, like *Mytilus californianus*, or cultures of *Gonyaulax catenella* gave saxitoxin. With dinoflagellates of the genera *Gonyaulax* and *Protogonyaulax* complex mixtures of closely related toxins were obtained. These toxic mixtures, varying in composition from strain to strain, can be traced back the strain of origin by indicating where the dinoflagellate was collected. Thus, the toxin mixtures called "tamarensis complexes" are further specified by indicating Ipswich (from the massive red tide of 1972), Perch Pond, or Mill Pond as the spots where the productive organism, *Gonyaulax tamarensis*, was collected along the Massachusetts coast.

These salt-like toxins have the property of permeating the membrane channels through which the cell normally imports nutrients, which leads to the rapid death of the organism. Such processes are deliberately induced in physiology laboratories, making these toxins important tools for physiological research. This is but an example of many toxic terrestrial and marine natural products that are used in research aimed at understanding our physiology better and to prevent diseases and alleviate our ailments in the long run.

Gymnodinium breve [= *Ptychodiscus brevis*], order Gymnodiniales, which was responsible for red tides along the Florida coasts which were accompanied by massive fish dying and human intoxication, is placed in the second group of toxic dinoflagellates. The toxins it produces, called brevetoxins, do not, as far as we know, accumulate in the body of any

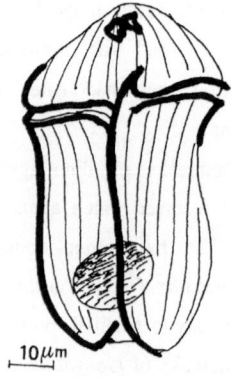

10 μm

Gymnodinium sp.

22

organism, which makes these toxins more difficult to study than the gonyautoxins. In order to obtain small amounts of brevetoxins to study their molecular structure, cultures of a few thousand liters, which were very expensive and difficult to effect, were needed.

One of the fish in Florida waters, called menhaden and belonging to the herring group, is resistant to the brevetoxins of *Gymnodinium breve* on which it feeds. The toxins are only found in the gut of the fish, which, once eviscerated, is safe for man to eat. This is not the case with another mammal, the bottle-nose dolphin, which has recently suffered mass mortality in Florida waters, not because of pathogenic microorganisms or viruses, but because it eats menhaden whole.

Brevetoxins are structurally quite different from gonyautoxins, lacking any salt-like character in particular, and affect cellular membrane channels which are needed to import food. Brevetoxins are acetogenins, but are far more complex in structure than EPA and vegetable oils.

The problem of detecting brevetoxins at the subtrace level with which they occur in fish and mollusks is a serious one as normal methods of analytical chemistry are not sensitive enough. However, in combining this with biological methods a solution has been offered to this problem: tritium-labeled radioactive precursors are fed to *G. breve* in culture whereby tritium-labeled brevetoxins are obtained. Labeled brevetoxins are then chemically linked to the protein bovine serum albumin to form antigens. This gives rise to determining brevetoxins with the high sensitivity of the antibody technique. This indirect method of detection is the radioimmunoassay method which has revolutionized the analysis of natural products in biological systems and is used daily in hospitals to evaluate the content of hormones in body fluids. This notwithstanding, chemists and biochemists, far from being satisfied, are actively searching for even more sensitive analytical techniques to carry out such determinations.

MARINE PRIMARY PRODUCERS

THE SPANISH CONQUISTADORES AND CIGUATERA POISONING

Gambierdiscus toxicus (Peridiniales), *Prorocentrum lima* (Prorocentrales), and *Amphidinium* sp. are placed in the third group of toxic dinoflagellates. *G. toxicus*, which was named after the ciguateric area of the Gambier Islands in tropical Pacific, is a benthic species which does not form blooms. However, it divides at an elevated rate during certain periods, settling preferentially on seaweeds. Outbreaks of this and other dinoflagellates intoxicate otherwise edible mollusc and fish in the Caribbean and in tropical Indo-Pacific areas. This is known as ciguatera poisoning.

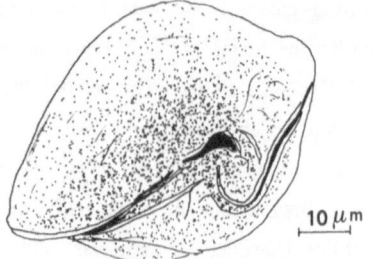

Gambierdiscus toxicus

Ciguatera is a name of Spanish origin that recalls the fact that the Spanish conquistadores in the Caribbean were intoxicated by the mollusc *Turbo pica* ("cigua" in Spanish). The responsible toxin has been called ciguatoxin as a result.

Ciguateric areas are: the south Pacific, central Indian Ocean, tropical American Atlantic, and the Coral Sea. On the Fiji Islands nearly all cases of marine food-poisoning are due to ciguatoxin. Recorded annual incidence of intoxication from 1975-1983 was 50 individuals per 100,000, which is typical for the south Pacific. In Miami the incidence is ten times higher. In Japan ciguatera is endemic only in southern areas (Okinawa, Amami) while in northern areas the ciguatera problem only arises if the fish are caught in the south Pacific.

In French Melanesia ciguatera is called "la gratte" because of a severe skin irritation which induces those affected to scratch themselves all the time. Though, following respiratory failure, fatal intoxications may occur, the illness usually only lasts for ten days and has much less serious immediate consequences than saxitoxin, gonyautoxins, or brevetoxins. This is not due to lack of potency but to the extremely low concentration of ciguatoxin in ciguateric fish. Ciguatoxin has additive effects, however. Thus, people who have recently suffered from ciguatera are prone to contract the illness again from a much smaller dose of the toxin. When one has suffered from ciguatera more than once a year, one should cease to eat fish, even from non-ciguateric areas, altogether for a while.

In folk medicine plant remedies are used to combat ciguatera. The most efficient remedy I saw was that of the Vanuatu, where the secret is held by an old woman who has always refused to sell the recipe to French scientist from the center of studies in Tahiti.

The economical consequences of ciguatera are quite serious, as I found out in February 1988 in New Caledonia, which is a beautiful large island of the Coral Sea discovered by James Cook in 1774 and colonized by the French in 1853, under Napoleon III. Economic resources were more scarce than usual during my visit due to the hurricane that had raged in January. Under such circumstances, lack of access to natural resources such as fishing is a serious drawback.

The demand for professional qualification and the desire to be independent prompted the foundation of a university; thus, l' Universitè Française du Pacifique has started courses in law and science in Noumea in cooperation with Tahiti, which lies several thousand kilometers further afield. In New Caledonia one hopes to attract Fijians; otherwise with a population of 130,000 how could the university survive in this area of the Pacific?

Ciguatoxin has been a very challenging scientific problem. It differs from the other dinoflagellate toxins in acting slowly, though it seems to act faster in New Caledonia than in Hawaii, possibly as a result of a higher concentration of the toxin. It proved difficult to cultivate *Gambierdiscus toxicus* and other dinoflagellates implicated in ciguatera poisoning, so that ciguatoxin has been extracted from intoxicated fish, mainly the moray eel (*Gymnothorax javanicus*), and in extremely small amounts. A less polar analog predominates in wild *G. toxicus*, as recently found in a cooperative effort among

French and Japanese scientists and the molecular structure has been elucidated: it is similar to that of brevetoxin-B.

DINOFLAGELLATES, SPONGES, AND WORMS: WHEN SYMBIOSIS IS USEFUL TO MEDICINE

Another toxic dinoflagellate of the third group, *Prorocentrum lima*, produces an antitumoral organic acid. This acid, which unfortunately is toxic for humans, was first found in the sponge *Halichondria okadai* that grows in Japanese waters (other than in the sponge *Halichondria melanodocia* found in the Florida Keys) and was therefore called okadaic acid.

The isolation of a compound such as okadaic acid from a sponge, while a microorganism is responsible for its *de novo* synthesis, is not an isolated case. There is a growing awareness that many natural products in marine animals are microbial in origin. Such microorganisms as the zooxanthellae, the epi- and endobacteria, and fungi, belong to

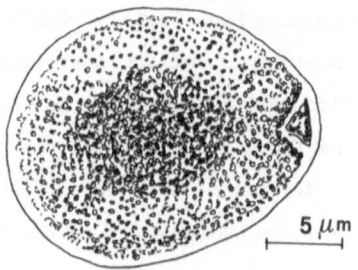

Prorocentrum sp.

groups that produce secondary metabolites and are widely distributed in marine animals. To reinforce this point, many of the metabolites isolated from marine animals belong to the same chemical classes as the typical metabolites of microorganisms.

A compound which is structurally similar to okadaic acid, acanthifolicin, has been isolated from the marine sponge *Pandaros acanthifolium* collected in the Virgin Islands. It is strongly cytotoxic and its origin is probably microbial though efforts to isolate acanthifolicin from cultures of microorganisms derived from *P. acanthifolium* have been frustrating. Such a failure should not be too surprising or discouraging, however. A precise diet is required by microorganisms to produce secondary metabolites; it is therefore possible that productive microorganisms of this sponge were discarded simply because an appropriate culture medium had not been found.

FURTHER DINOFLAGELLATES AND CHRYPTOPHYCEAE OF MEDICAL INTEREST

Another dinoflagellate of the third group, *Amphidinium* sp., is an intracellular symbiont of the small green flatworm *Amphiscolops* sp. present in Okinawan waters. This dinoflagellate in culture gave a mixture of three acetogeninic substances, the amphidinolides, which were separated from one another and proved highly cytotoxic with leukemic cells, particularly amphidinolide-C.

A bloom-forming dinoflagellate from La Parguera, Puerto Rico, *Goniodoma pseudogoniaulax*, contains an antifungal substance, goniodomin, the structure of which is unknown. At a later stage a definite substance, called goniodomin-A, which is active against the man-pathogenic yeast *Candida albicans* and has ichthyotoxic activity, has been isolated from blooms of *Alexandrium hiranoi* [= *G. pseudogonyaulax*] in Japan; it is a complex acetogenin. The amphidinolides and goniodomin-A belong to the macrolide class, as does sphinxolide, which was isolated from an unidentified nudibranch mollusc collected in Hawaiian waters.

Hormothamnione, which is highly toxic to leukemic cells, was isolated from a "yellow slime" (collected in shallow reefs along the northern coast of Puerto Rico) which was thought to be a senescent mass of the cyanobacterium *Hormothamnion enteromorphoides*. Subsequently, this "yellow slime" was identified to be *Chrysophaeum taylori*, which is a brown alga belonging to the small group of marine Chryptophyceae. Hormothamnione is worth particularly paying attention to in that, as a simple aromatic molecule, it is easy to synthesize. The total synthesis of marine natural products, solicited by both academic and economic interests, is an area of current rapid development.

DINOFLAGELLATE STEROLS

Social relevance has made dinoflagellates the subject of further study, revealing that their metabolism is aimed at producing various acetogenins and acetogenin-sugar complexes and not merely sugars, which is often the case with marine organisms.

Dinoflagellates also produce steroids, a name that brings to mind three problems: the contraceptive pill, elevated cholesterol level in the blood,

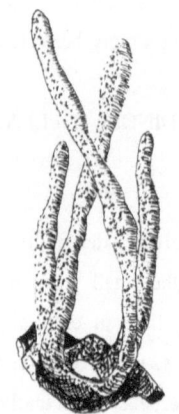

Briareum asbestinum

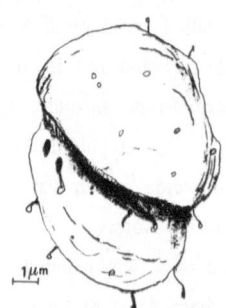

Zooxanthella microadriatica

and "doping" in athletics. Actually, some steroids exert a number of normal functions in our bodies and in other living organisms. Dinoflagellate steroids have received particular attention in relation to their association with marine invertebrates, typically reef corals and sponges. These studies have revealed that the main sterol of gorgonians, such as *Briareum asbestinum* from the Caribbean, is dinosterol, which is a steroid atypical to higher animals, where the main sterol is cholesterol. In fact, *B. asbestinum* is unable to synthesize dinosterol and receives it from zooxantellae hosts; this is proved by the fact that cultures of the dinoflagellate *Zooxanthella microadriatica*, harbored by *B. asbestinum*, produce dinosterol[4].

---

[4]Some scientists prefer the name *Symbiodinium* for the algal genus *Zooxanthella*.

PHYTOPLANKTON

This is a further example of intricate metabolic pathways that result from the ubiquity of associations among living organisms in the sea. In the absence of long and complex studies (such as those described above for *B. asbestinum* and *Z. microadriatica*) it is difficult to build up a map of representative metabolites for the various groups of marine animals.

Dinosterol is a chemical mark for both living dinoflagellates and sedimentary material of dinoflagellate origin. In fact dinosterol is present in most dinoflagellates while it is absent from other marine and lacustrine algae.

Phytoplankton is also rich in unusual fats.

CYANOBACTERIA AND EVOLUTION

Cyanobacteria represent an evolutionary stage in the development from primitive prokaryotes to eukaryotes. Unicellular forms are closer to prokaryotes, whereas filamentous and gliding forms are more evolved. With respect to typical bacteria, cyanobacteria are more elaborate cells which are capable of photosynthesis either without oxygen evolution, like bacteria, or with oxygen evolution, like higher plants and algae. Cyanobacteria are also capable of oxidative reactions that lead to organic acids which, like acetic acid, are the building blocks for constructing acetogenins. Cyanobacteria may thus be considered aerobic bacteria.

Because of their very old age and the properties mentioned above, cyanobacteria may have been responsible for the appearance of molecular oxygen in the world. Moreover, analogous to zooxanthellae-coral symbiosis, cyanobacteria play the important role of transferring photosynthetic nutrients to sponges in coral reefs.

Many cyanobacteria are also capable of aerial nitrogen fixation. To acquire this property, the cyanobacteria had to overcome the extreme lability of their nitrogen-fixing enzyme (nitrogenase) toward molecular oxygen. Two strategies were adopted by the cyanobacteria for this purpose. Some cyanobacteria have specialized organelles where nitrogen fixation, but no oxygenic photosynthesis, takes place. Other cyanobacteria carry out nitrogen fixation at night when photosynthesis is impossible. In either case, the nitrogenase is prevented from coming into contact with molecular oxygen. Cyanobacterial nitrogen fixation is used for growing plants, e.g. rice, which would otherwise require nitrogen fertilizers.

Some cyanobacteria are also capable of degrading heavy pollutants such as aromatic hydrocarbons and amines. Although certain algae have a similar capacity, cyanobacteria have the advantage of requiring little molecular oxygen and low light intensity to grow, promising a new approach to waste conversion. However, agricultural applications require pesticide-resistant strains of cyanobacteria, since pesticides may have an adverse effect on the growth of non-resistant cyanobacteria. As Nature has only recently been confronted with synthetic pesticides, and is slow to adapt, faced with the problem of maintaining an overall balance, pesticide-resistant natural strains of cyanobacteria rarely occur. Today they are obtained by treating natural strains of cyanobacteria with mutagenic compounds, such as nitrosoamines. This raises many problems, the most obvious of which is how to control strains of pesticide-resistant cyanobacteria. I myself believe that we should reflect on whether it is appropriate to continue trying to "improve" on natural cyanobacteria by inducing quick mutations.

GAIA, PHYTOPLANKTON, AND CLIMATE

The fact that living things influence their surroundings is an everyday experience and cyanobacteria are a prominent case, if they have really been responsible for the appearance of molecular oxygen on Earth. This and similar cases are contemplated by Gaia, a theory named after the Greek goddess of the Earth. According to this theory by Lovelock, life is the dominant force in the course of the Earth which is seen as a superorganism that manages to achieve and maintain the best conditions for life.

Now that Gaia has been refuted by her creator, those involved with environmental natural gases have proposed a more limited version of this theory. Thus, if the concentration of carbon dioxide (a greenhouse gas that helps to keep the Earth warm) exceeds the present level any further, melting of land-based ice sheets and glaciers will cause the level of the oceans to rise at a rate in excess of the presently calculated 2.4 millimeters per year and finally the oceans will start to boil under the action of the sun. They also point out that corals and mollusks fix atmospheric carbon dioxide into calcareous materials and that photosynthetic organisms, while releasing molecular oxygen, transform carbon dioxide into molecules which are essential to the survival of all organisms. Finally, they observe that diatoms produce a small sulfurated compound, called dimethyl sulfide, which, as we already know from dinoflagellates and penguins, results, together with acrylic acid, from the decomposition of ß-dimethylpropiotethin. Dimethyl sulfide escapes into the air, where it undergoes oxidation to sulfate, thus generating clouds over the sea that reflect the sun rays and prevent the oceanic waters from becoming too warm.

Other scientists object to these views, however, reasoning on sulfur dioxide, which is another gaseous precursor of atmospheric sulfate. According to them, sulfur dioxide, as a man-made gas from fossil fuel combustion, has a higher concentration in the northern than in the southern hemisphere. Therefore, assuming that the above "cloud effect" is really operative, one would expect it to be particularly intense in northern areas of the world. As this is not the case, these scientists even refuse to accept the smoother version of Gaia.

In this tricky matter we, the spectators, acknowledge that physical processes are accompanied by living processes on Earth, which would be quite a different planet in the absence of life.

THE SWIMMER'S ITCH

Pharmacological interest concerning cyanobacteria dates back as early as 1500 B.C. when *Nostoc* species were used to treat gout, fistula and cancer, although I do not know how successfully. In any event, *Nostoc* species were highly praised even in the Middle Ages.

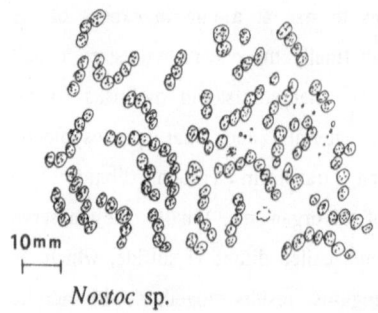

10mm

Nostoc sp.

Modern research has revealed that one of the most widespread cyanobacteria in tropical and subtropical waters, the filamentous *Lyngbya majuscula* [= *Microleus lyngbyaceus*], contains two potent inflammatory products, aplysiatoxin and debromoaplysiatoxin, which cause serious contact dermatitis, known as "swimmers' itch" in Hawaii

These toxins were first found in tiny amounts in the sea hare *Stylocheilus longicauda* which inhabits the same waters as *L. majuscula*. This sea hare is so unusually small in size that 5000 specimens were needed to extract a few milligrams of the toxin.

The ancient Romans were already aware of the toxicity of sea hares. Thus, Pliny the Elder fabulously described that if a pregnant woman saw such mollusks she would feel nauseous and miscarry. However, in our times the aplysiatoxins have been isolated from *S. longicauda* in response to biotests which revealed inflammatory activity on the human skin.

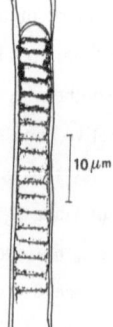

10 μm

Lyngbya majuscula

The mollusc rather than being injured is protected by cyanobacterial toxins; lacking a shell and being only able to swim slowly, it would make easy prey without these toxins.

CYANOBACTERIA AS DEVILS

Another metabolite of *L. majuscula*, the alkaloid lyngbyatoxin-A, is not only a very potent inflammatory product but also an even more harmful tumor promoter. Danger is not limited to the marine environment, as certain bacteria of the genus *Streptomyces*, which live on terrestrial soil, produce similar metabolites with the same horrible effects.

Freshwater cyanobacteria may also be dangerous, like *Microcystis aeruginosa*, which belongs to the order Chroococcales and is a more primitive species than filamentous cyanobacteria such as *Lyngbya majuscula*. Studies of *M. aeruginosa* have led to the characterization of polypeptidic toxins. Yet another harmful freshwater cyanobacterium is *Anabaena flos-aquae*, which produces fast-death alkaloids.

From these stories one might conclude that cyanobacteria are devils designed by Nature to offend man; a surprising thought for those who believe that everything that is natural must be good. Nevertheless one should not blame Nature; once again one should remember that in her unbiased overall balance, Nature can not indulge man.

3 Seaweeds, seagrasses, and emergent plants

Seaweeds are macroscopic algae that can be classified in three divisions, the red, the brown, and the green seaweeds. They inhabit rocky marine areas in shallow waters and although a few species of macroscopic algae are also known to exist in freshwater, they never attain the large size and population density of certain seaweeds.

The phylogeny of the red seaweeds is markedly different from that of the green seaweeds, while the ancestral precursor of the brown seaweeds is still unknown. The evolutionary stage in each of these three divisions also differs: the green seaweeds are the most evolved as they lie closer to the higher plants in the evolutionary scale, at least as far as we can judge from the distribution of natural products.

Seagrasses are flowering plants which, together with many freshwater herbs and the palms belong to the class Monocotyledonae. Despite a limited number of species, 48 in 12 genera, they have colonized nearly all sandy coastal areas.

Marine emergent plants are higher plants which inhabit the intertidal shore.

SEAWEEDS AND ZONATION OF COASTS AND OCEANS

Seaweeds dominate rocky areas and can also be found in sandy zones, provided that there are stones that are the support on which they grow. Epiphytic or parasitic species, which are very common, grow on seagrasses or on other seaweeds. Overall, the contribution of seaweeds to the primary productivity in coastal areas is generally locally dependent and may surpass seagrasses in certain rocky areas.

Because they lack true roots, seaweeds need a hard substrate on which to set; only a few tropical species have root-like appendices and can therefore grow on soft substrates. In no case, however, can seaweeds draw nutrients from the substrate, unlike terrestrial plants and seagrasses.

In some rare cases, notably *Sargassum vulgare* and *S. filipendulum*, seaweeds have adapted to planktonic life; their drifting on the Sargasso Sea and traveling up the Caribbean coast is a phenomenon that has fascinated man since the days of Columbus.

Habitat preferences differ from group to group. Red seaweeds can be found at all latitudes, brown seaweed dominate in temperate and cold waters, and green seaweeds in the tropics. Vertical distribution depends a lot on local conditions and particularly on the transparency and the temperature of the waters. In order to gather enough light for photosynthesis, seaweeds belonging to different divisions have developed different accessory pigments according to the change of color with the depth of the water: at depths of 5-10 meters turbid coastal waters are greenish whereas oceanic waters, or particularly clear coastal waters, are bluish. Red seaweeds are capable of absorbing light of both these colors, so that they can live at any depth provided that there is enough light overall. In contrast, green seaweeds do not absorb green light, so that in turbid waters they can only be found at a shallow depth. Distribution is different in clear waters, such as in the tropics, where green algae are found at all depths. Brown seaweeds have an intermediate light-absorption behavior so that it is more the temperature of the waters that determines both their latitudinal and vertical distribution rather than the depth. As a rule, brown seaweeds prefer cold waters.

Sargassum filipendulum

MARINE PRIMARY PRODUCERS

In any event, the turbidity of the water limits the depth at which the seaweeds can grow. In northern Europe the lower limit is 30 meters whereas in the tropics 100 meters is typical. Occasionally, during explorations with a submersible, red algae have been spotted at a depth of 280 meters. However, in the tropics it is both the shortage of nutrients and heavy grazing that inhibit the overgrowth of seaweeds. In conclusion, except when limiting the geographical area, no precise vertical distribution of seaweeds can be given.

Both this and any subsequent discussion on the vertical distribution of marine organisms are made clearer if we consider the zonation of coasts and oceans. The sea is primarily divided into the benthic zone (the bottom) and the pelagic zone (the open waters). The benthic zone is subdivided into the supralittoral, the littoral (or intertidal zone, which is subjected to tides), the sublittoral, the bathyal, the abyssal, and the hadal zone.

The sublittoral zone extends from the tidal zone to about 250 meters in depth. It is impressive that the abyssal zone, at mean depths below 2.5 kilometers, accounts for about 85% of the sea floor, and thus nearly 60% of the Earth's surface. This results in an average depth of the sea of about 4 kilometers.

The pelagic zone is subdivided into the neritic zone, which is above the continental shelf, and the oceanic zone, which is subdivided further into the epipelagic, mesopelagic, bathypelagic and abyssopelagic zone.

Tides are of great importance. Whereas in open oceanic waters tides never exceed one meter, in coastal waters the tidal range depends on local conditions. Thus, tides pass unnoticed in most Mediterranean coasts, whereas in certain bays (like the Bay of Fundy in Canada) they can go beyond 20 meters. Tidal zones are areas of stress that strongly influence the vegetation and the coastal distribution of invertebrates.

The size of seaweeds spans from some tiny epiphytic or parasitic species, which measure a few millimeters in length, to giant brown seaweeds of the order Laminariales or Fucales (kelp) the ashes of which are used as a source of iodine. A striking case is *Macrocystis pyrifera*, of American Pacific coasts, the arms of which extend to over 50 meters.

With regard to algae even the coral reef is a peculiar environment. Though most of the carbonate mass of coral reefs is fixed by corals (particularly scleractinians), both green, like *Halimeda tuna*, and red calcareous algae, of the genera *Lithothamnion* and *Lithophyllum*, are important reef builders.

Macrocystis pyrifera

THE BIG BUSINESS OF SEAWEEDS

Indigenous relics have demonstrated that seaweeds were already being eaten in the Japanese Archipelago ten thousand years ago. This tradition has continued to our day and has also spread China and other countries, where, like in Japan, seaweeds are of enormous nutritional importance and are cultured for this purpose.

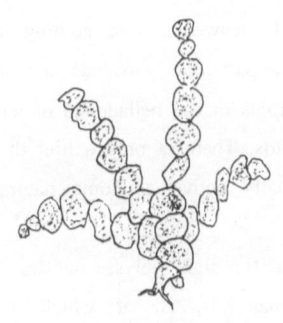

Halimeda sp.

Nutritionally speaking we can not expect too much from seaweeds, however. They are vegetables and nothing more. In general, edible seaweeds are rich in sugars, vitamins, and those trace elements which are required in the human diet. With the exception of the red seaweed *Porphyra*, which will be discussed later, and a few other cases, seaweeds are scarce in proteins and the balance of amino acids needed by man is not as appropriate as that in meat and eggs.

37

Algal products are also of great economic importance, principally polymeric carbohydrates, called polysaccharides, which have the role of reserve or structural materials and which resemble polysaccharides in higher plants. In addition, seaweeds have sulfated polysaccharides with properties that are similar to the pharmaceutically important heparins which are mammalian sulfated polysaccharides without a counterpart in higher plants.

The reserve polysaccharides of seaweeds are divided into two categories, starch-like and laminaran. Algal starches are similar to plant starches but have a lower molecular weight, viscosity and capacity of binding iodine.

Cellulose is a structural polysaccharide common to all seaweeds and resembles the cellulose in higher plants. In addition, and this is of great economic value, brown seaweeds have alginic acids and alginates and red seaweeds agar and carrageenans as structural polysaccharides.

The pharmaceutical use of algae was already contemplated in the first Chinese pharmacopeia, the "Pen-Tso Chin", which probably was written in the first or second century B.C., as well as in the Ayurveda, the multimillenary Indian system of medicine. However, these rather fabulous reports are impregnated with superstition: Voltaire had not yet arrived to replace superstition with rationality.

Actually, the place of the seaweeds in pharmacy is still very limited, except where real or putative beauty products are concerned. Seaweeds have nothing in common with bioactive molds and terrestrial plants; in particular seaweeds lack in general physiologically active products, such as the opioids or the belladonna of terrestrial higher plants, as well as the antibiotics of molds. There is only a hint that certain algal polysaccharides may be as physiologically active as some natural products extracted from higher terrestrial plants.

Currently the most successful pharmaceutical use for algal polysaccharides is as wound healer, anticoagulants, and supporting materials, all of which are unrivalled. Another important aspect concerns poor countries where the soil is deprived of salts by frequent rains. Eating seaweeds, particularly brown seaweeds which contain large amounts of iodides, can relieve the problem of endemic goitre, which is a disorder that still affects a large part of the population in such areas.

SEAWEEDS AND WORLD POLLUTION

Common species of seaweeds, belonging to all three divisions, release small halogenated hydrocarbons identical or similar to typical industrial pollutants. The seaweeds responsible belong to the genera *Fucus* (serrated-, flat-, and bladder-wrack) and *Ascophyllum* (knotted-wrack) in the brown, *Gigartina* and *Asparagopsis* in the red, and *Enteromorpha* (green weed) and *Ulva* (sea lettuce) in the green seaweeds. Such seaweeds produce halogenated derivatives of the Earth's gases (methane, propane, and butane with hydrogen atoms

Gigartina sp.

replaced by chlorine or bromine). With such seaweeds the content of halogenated hydrocarbons is impressively high (up to 0.1 grams per gram of dried seaweed)

Asparagopsis armata

which means that the total amount of such products released by seaweeds is comparable to that escaped from chemical plants.

Such halogenated hydrocarbons, some of which have a pleasant odor, are considered harmful to man; which, once again, leads us to conclude that Nature can not indulge man. However, naturally-occurring fluorocarbons do not exist and in general natural organic compounds of fluorine are rare. Man is responsible for the production of fluorochlorocarbons which serve as propellants for shaving creams and perfumes or as cooling liquids and which are said to damage the ozone layer. Even terrestrial plants that are capable of absorbing nitrogen oxides from exhaust fumes are inefficient where fluorochlorocarbons are concerned. However, I am puzzled by journalists (and scientists) who spend so many words on fluorochlorocarbons and forget about the extremely dangerous nitrogen oxides which are released *in situ* by jets.

PHOTOSYNTHETIC ACCESSORY PIGMENTS OF RED SEAWEEDS: A PHYLOGENETIC MARK

Seaweeds, in particular red seaweeds which derive from cyanobacteria, have inhabited the sea for a long time. The red seaweeds have inherited much of the photosynthetic apparatus of their ancestral precursors; their plastids have the pigments of cyanobacteria instead of the light-gathering chlorophyll-b of green seaweeds and higher plants. These pigments, which are structurally related to chlorophylls and the prosthetic group of heme and cytochromes, are a phylogenetic mark.

This is not an isolated example: biochemical indications of phylogeny are much sought to corroborate and to extend classical morphological and embryological indications.

RED SEAWEEDS AS FOOD: THE FIRST COMPONENT OF THE BIG BUSINESS AND OGONORI POISONING

Some red seaweeds are made into highly appreciated nutritional products worldwide and not only in Japan, where the genus *Porphyra* is the most intensely cultured, particularly for storable products (nori). Unusually, these seaweeds are quite rich in proteins, up 30% of their dry weight. Interest in these products is so high that American "nori" is prepared from local *Porphyra nereocystis* and is marketed in the USA.

SEAWEEDS

Being aware of so many poisonous terrestrial plants, one asks oneself about the toxicology of seaweeds. In general seaweeds are considered to be harmless to man, a conclusion that is amply justified. Exceptions of which one has known for a long time are certain green seaweeds belonging to the genus *Caulerpa*, to be discussed later in this book, which may induce unpleasant physiological effects when ingested.

Recently, however, there have been mortal intoxications in Japan, which were caused by red seaweeds belonging to the genus *Gracilaria* which are in general scarcely appreciated for nutritional purposes; they are mostly used for industrial purposes, though the species *chorda, compressa, confervoides, cornea, coronopifolia, edulis, eucheumoides, taenioides,* and *verrucosa* are also eaten. It would seem that the victims had eaten *G. chorda* and *G. verrucosa*, the Japanese "ogonori", from which the term "ogonori poisoning" was coined.

Prior to being eaten, these seaweeds had been soaked overnight in freshwater to remove abundant inorganic salts; therefore any poisoning due to inorganic salts was ruled out. It was then discovered that the poisonous agent of *G. chorda* and *G. verrucosa* is PGE_2, a prostaglandin which was also discovered recently in another seaweed of the same genus, *Gracilaria lichenoides*, collected in East Australian waters. Prostaglandins were at that time already known to be involved in a number of physiological processes with mammals, such as inflammation of tissues, and to be contained in gorgonians, such as *Plexaura homomalla* which is found in the Caribbean. However, the presence of prostaglandins in a vegetable, like *Gracilaria*, was unprecedented.

AGAR-AGAR AND OTHER POLYMERS OF THE RED SEAWEEDS: THE SECOND COMPONENT OF THE BIG BUSINESS

Most polysaccharides extracted from red seaweeds form gelatins which are called phycocolloids and are used by the food, cosmetic, and pharmaceutical industry. They

41

are mainly used as emulsifying, binding, suspending, and thickening agents, but also as degradable vehicles for drugs. They are also used in the leather, paper, and textile industry, and as a substrate for growing microorganisms.

The widespread use of phycocolloids reflects the fact that their properties can not be easily reproduced by synthetic polymers, at least not for the same price at which phycocolloids can be obtained from Nature.

The best known phycocolloids are agar and the carrageenans. Agar, which is the oldest and formerly known as agar-agar, a name Malayan in origin, is a mixture of polysaccharides with a differing molecular weight and consist mainly of neutral agarose, pyruvated agarose, and a sulfated polymer of galactose. Agar is industrially subjected to fractionation in order to prepare its component polymers; of these, agarose is largely used in immunological studies. Agar is mainly prepared from red seaweeds that belong to the genera *Gelidium*, *Gracilaria*, *Acanthopeltis*, *Ahnfeltia*, and *Pterocladia* in East Asia (Japan, URSS, Korea, India), America (USA, Chile), Europe (Spain, Portugal), and Africa (Morocco).

The carrageenans, a name derived from the town Carragheen in Ireland, were already used as laxatives by Romans. Main producers are the USA, France, Great Britain, and Norway from *Chondrus crispus*, *Gigartina stellata* (which are harvested collectively and called "Irish moss"), *Halymenia venusta*, *Laurencia papillosa*, *Sarconema filiformis*, and various species of *Hypnea* and *Eucheuma*.

Apart from their extended use as gelling, emulsifying, and stabilizing agents for food and non-foods, and in particular for milky and waterish media, the carrageenans are also used to stimulate the growth of connective tissues and to lower the level of cholesterol in the blood. In this aspect the carrageenans surpass all other phycocolloids. They are therefore likely candidates for antiatherosclerotic drugs, provided that long treatments do not deprive the organism of the necessary salts which may be coordinated and rendered insoluble by the polysaccharide. This problem will be reconsidered later with the alginates of brown seaweeds.

Similar polysaccharides are also commercialized, e.g. Ceylon moss, Chinese moss, Furcellaran, also named Danish Agar, Gracilaria Gum, Gulaman, and Phyllophoran (in the food industry), Ginnansô (as an adhesive in East Asia), and Iridophycan (to prevent blood coagulation). Global algal extracts, such as Funoran, are used in cosmetics.

A different product is Maërl, which consists of the calcified remains of the red calcareous algae *Lithothamnion corallioides* and *Lithothamnion calcareum*. It can be collected in abundance in Brittany waters and it is used in the pharmaceutical and cosmetic industry as well as for special sewage processes, particularly radioactive

Sargassum vulgare

materials, due to its strong absorbing properties. We have found that these algae contain unique lipids of the family of arachidonic acid.

RED SEAWEEDS IN PHARMACY AND IN NEUROBIOLOGY

Besides being used as food and a source of agar and carrageenans, some seaweeds have anthelmintic properties. Intestinal worms are still a common problem,

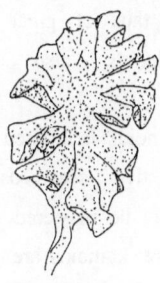

Ulva lactuca

particularly in central Africa; the relatively harmless worms *Ascaris lumbricoides* and *Oxyuris*, but also *Taenia* and *Trichuris*, not only subtract nutrients from the organism, but also produce substances that inhibit enzymes and as a result weaken the capacity to absorb foods. Common anthelmintic seaweeds are *Sargassum vulgare, Ulva lactuca, Durvillea antarctica*

43

(used in New Zealand by the Maoris), and *Alsidium helminthocorton*. The latter, called Corsican moss, was already used by the Greeks and in the eighteenth century has replaced the less efficient red seaweeds *Corallina officinalis*, *Jania rubens*, and *Rhodymenia palmata* in the Mediterranean. *Hypnea musciformis* also has anthelminthic properties but it is mostly used for the production of carrageenans.

The most active anthelminthic seaweeds are *Digenea simplex* and *Chondria armata*, which are by no means widely spread, however. The active principles, α-kainic acid in the former and domoic acid in the latter, are both non-protein amino acids. α-Santonin, which is an anthelmintic terpene extracted from terrestrial plants, has a synergistic effect

Jania rubens

with α-kainic acid; a mixture of the two compounds is commercially available and doses of about 20 milligrams per day over a period of some weeks will eliminate intestinal worms.

Digenea simplex

α-Kainic, domoic acid, and related amino acids, which are collectively called kainoids, also reveal important neuroexcitant properties and are thus of great use in neurobiology. Kainoids also possess insecticidal properties.

The use of kainoids is limited by the scarcity of productive seaweeds. This problem can probably be mastered by total synthesis in the future: kainoids are simple molecules, so that practical total syntheses have already been devised. This will free us from our dependence on Nature to supply us with kainoids, although a crisis for the extractive industry of these substances is foreseeable. This recalls a case concerning the insecticidal pyrethroids of the terrestrial herb *Chrysanthemum cinerarifolium* 70 years ago. Efficient, total syntheses of the pyrethroids were devised on a large scale, which was the cause of a harsh economical crisis in Dalmatia, a Yugoslavian territory where a large part of the population lived from extensive cultures of *C. cinerarifolium*. Curiously, Leopold Ruzicka, the chemist who invented these syntheses in Zürich, was of Yugoslavian extraction.

Natural products with cellular activity, which are of great interest to the pharmaceutical industry at present, have also been isolated from seaweeds. Both *Laurencia thyrsifera* (Hook) from New Zealand and a variety of *Laurencia obtusa* from Teuri Island, Japan, contain a terpene, called thyrsiferol, which is highly cytotoxic in vitro to abnormal cells.

Enzymatic activities are also known with seaweeds. Thus, the red seaweed *Ptilota filicina* (Ceramiales), collected along the coasts of Oregon, contains a substance, structurally similar to the polyunsaturated fatty acid EPA of the diatoms, which inhibits the enzyme Na^+,K^+-ATPase.

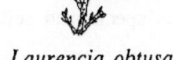

Laurencia obtusa

CHEMOTAXONOMY WITH RED SEAWEEDS

In all three divisions of seaweed the prevalent amino acids are alanine, aspartic acid, and glutamic acid. You should have heard of the latter at least, which is listed as a food additive on soup cubes. All three amino acids are needed for the biosynthesis of proteins by animals. Not all seaweeds follow this pattern, however. Certain seaweeds of the Rhodomelaceae family, order Ceramiales, have a high content of the protein-amino-acid proline, which is a chemical mark of these seaweeds.

45

MARINE PRIMARY PRODUCERS

This is an example of natural products which help to classify living organisms. The concordance between biological classification, which is based on morphological, phylogenetical, and embryological observations, and the distribution of natural products, must reflect genetic control. This field is called chemotaxonomy, a borderline area between biology and natural product chemistry which has a long tradition in terrestrial botany.

CHEMICAL ECOLOGY OF RED SEAWEEDS

Certain odoriferous products of the red seaweeds play an important role in the regulation of marine life. They are abundantly produced by seaweeds of the genera *Plocamium*, *Chondrococcus*, and *Ochtodes*, which belong to the order Gigartinales and are commonly found in tropical and subtropical Pacific areas of heavy surf; on crushing such seaweeds a pleasant, turpentine-like odor arises from oil droplets stored in specialized cells.

Plocamium sp.

These compounds, which are small terpenes, were first isolated from the sea

Aplysia californica

hare *Aplysia californica* which feeds on these seaweeds and retains the active principles mainly in the digestive gland as a protection from predators. This case parallels that of another sea hare, *Stylocheilus longicauda*, which is protected by aplysiatoxin, a product of the cyanophyta *Lyngbya majuscula*.

46

These odoriferous terpenes are also employed by the Gigartinales for the same purpose. In stress areas such as the tropics, the pressure to find food is so heavy that exposed organisms such as these seaweeds have had to develop a chemical defense mechanism in order to survive. Field tests have shown that odoriferous terpenes of *Ochtodes crockeri* from the Galápagos Islands have feeding inhibition properties down to the 100 ppm level against the damselfish *Pomacentrus coeruleus*. Since these odoriferous compounds make up 15,000 ppm of the dry weight of these seaweed, there is no doubt that these compounds exert a defensive role.

Certain red seaweeds produce compounds with a hormone-like activity. Thus, the metamorphosis of the larvae of the mollusc *Pecten maximus*, the delicious coquille Saint Jacques of Brittany, is induced by a simple compound, jacaranone, which is produced by the local red seaweed *Delesseria sanguinea*. The name of the compound (the carbon skeleton of which resembles that of the dienone portion of thelepin, which has been isolated from the annelid *Thelepus setosus*), is derived from the terrestrial plant *Jacaranda caucana* from which it was first isolated.

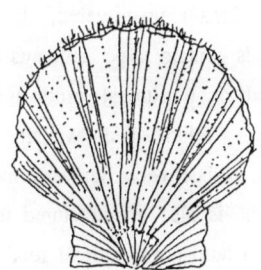

Pecten maximus

Another case of different functions exerted by the same compound is that of the amino acid GABA. This is a neurotransmitter with man and a settlement inducer (produced by calcareous red algae in the waters of Brittany) for the larvae of the mollusc abalone ("ormeau" in Brittany, where it is highly appreciated).

This is a further example of different biological functions exerted by the same natural product. It is a sign that variations in the molecular structure of natural products and their receptors can not be endless. Therefore, Nature has to use, or finds it economical to use the same key for various, not intercommunicating, doors.

We do not yet know the role nor have we found uses for the great variety of other metabolites produced by red seaweeds, mostly by species of the more evolved class, Florideophyceae. The work on these seaweeds has mainly served to better appreciate Nature as a chemical architect in the construction of complex molecules. The day when these discoveries will become useful to man may not be soon, but will surely come.

POLYMERS, FOOD, AND DRUGS FROM BROWN SEAWEEDS: THE THIRD COMPONENT OF THE BIG BUSINESS

Though certain basic aspects of brown seaweeds, first of all their phylogenetic origin, remain unanswered, these seaweeds have been largely exploited. Brown seaweeds contain larger amounts of iodine than red and green seaweeds and are used as food or nutritional products. Important nutritional products in East Asia are "kombu", "hijiki", and "wakame", prepared from *Laminaria*, *Hizikia*, and *Undaria* species, respectively. Seaweed meal, prepared in Norway mainly from *Ascophyllum nodosum*, is also manufactured in Japan, where it is used as manure in agriculture and as a human and animal food additive.

Polysaccharides from brown seaweeds are also important. The best known are alginic acids and their salts (alginates), which are prepared mainly from the giant kelp, *Macrocystis*, from the coasts of Pacific North America, New Zealand and Australia, and *Laminaria*, from the North Atlantic and Japanese coasts. These polysaccharides are made up of approximately 1800 units of D-mannuronic and L-guluronic acid. Major producers are the USA, France, Great Britain, Japan, Norway, and the USSR.

Alginates, such as sodium alginate, are used in much the same way as the polysaccharides of red seaweeds. Alginic acids are capable of absorbing up to 60% of their weight in salts, including strontium salts, which are not absorbed by the polysaccharides of red seaweeds. In this way strontium salts precipitate as insoluble

48

alginates, which are not absorbed by the intestine and are thus removed from body fluids. This can be useful for preventing substitution of calcium in our bones by radioactive strontium released by nuclear plants or bombs. Prolonged treatment with alginates also results in the strontium fixed in the bones to be replaced with calcium as the result of the equilibrium between fixed and dissolved strontium salts.

In this appealing scenario there are clouds, however: to what extent are alginates also capable of removing the elements which are needed by the human body, such as iron? Any risks have to be assessed in order to avoid anemia or other pernicious diseases, before introducing the practice of long treatments with polysaccharides.

In contrast, alginates can not remove lead from our organism and are thus incapable of detoxifying our children[5], who incorporate lead compounds released by cars into the atmosphere. Another phycocolloid of brown seaweeds, called fucoidan, might be of some assistance in this matter. Lead salts are absorbed by fucoidan and precipitated as insoluble complexes under the alkaline conditions of the intestine, but not under the acidic conditions of the stomach. Fucoidan also has such a potent anticoagulative effect on the blood that it surpasses the heparins as an antithrombotic agent.

There are not many non-polymeric compounds of brown seaweeds that have raised concrete interest as drugs so far. Perhaps the best known of these compounds is the non-protein amino acid laminine which can be found in various species of *Laminaria* and the edible *Heterochordaria abietina* [= *Analipus japonicus*]. This

---

[5]Sadly enough, fetal exposure to lead compounds (at the current level of polluted cities) is conductive to a lead level in the blood of babies sufficiently high to cause impairment in their mental development. This is only one of the risks of car pollution. Other major problems can not be alleviated with the use of lead-free gasoline, nor is the use of catalytic converters safe as they release extremely small solid particles, which are as dangerous as asbestos, into the air. We have been too rash in removing the old electric vehicles from our cities and in allowing the endothermic motor to take over.

amino acid has strong transitory blood-pressure lowering effects and (by blocking acetylcholine and histamine effects) muscle-relaxing properties.

Brown seaweeds have also evolved chemical defense systems. This is the case with *Stypopodium zonale* in the Caribbean, which secretes stypoldione, a mixed-biogenesis terpene with lethal action on herbivorous fish and inhibition of the first cleavage of fertilized sea urchin eggs. Along the same lines, *Cladosiphon okamuranus* (an edible seaweed of the order Chordariales which is cultivated in Japan) produces a fatty acid which inhibits the growth of both red tide phytoplankton and epiphytic red seaweeds.

ARSENIC AND BROWN SEAWEEDS

We are already aware that certain seaweeds accumulate various elements. While iodine, potassium, and the trace elements required in our diet make seaweeds a valuable food, one does wonder whether dangerous elements also accumulate. Unfortunately, the answer is affirmative. Thus, the kelp *Ecklonia radiata* has been recently found to accumulate substantial amounts of arsenic (about 10 milligrams per kilogram of wet kelp), mostly as water-soluble arsenosugars, such as 'eckloniaarsenoribofuranoside', in unpolluted waters of Western Australia. Even *Fucus* species accumulate arsenic, mainly as complexes which are soluble in our body fats. Luckily, none of the *Fucus* species nor *E. radiata* are eaten. However, other species of *Ecklonia* are used in the diet of certain human populations.

Arsenic can enter the food chain in other ways, too. For instance, in Shark Bay, Western Australia, arsenosugars quite similar to those of *E. radiata*, such as 'tridacnaarsenoribofuranoside', accumulate in the kidney of the giant clam, *Tridacna maxima*, which does not feed on either *Ecklonia* or *Fucus*. It is possible that symbiotic zooxanthellae are responsible for the incorporation of these arsenosugars.

The situation of the shrimp *Sergestes lucens*, and probably of all shrimps which feed on phytoplankton, is similar. This crustacean, obtained at the Tokyo

Central Wholesale Market, has recently proven to contain arsenobetaine, which is also found in fish and sharks.

We can not even rule out that arsenic, or other dangerous elements, are present in edible seaweeds and because of this, harvesting seaweeds in polluted areas is hazardous and is particularly serious in the case of radioactive compounds from natural sources or released by nuclear plants and bombs.

SEXUAL ATTRACTION WITH BROWN SEAWEEDS

Few people would have brown seaweeds in mind when considering sexual matters. Actually, these algae have developed a most efficient system of sexual attraction; they produce volatile compounds which have the property of coordinating cellular activities in sexual reproduction, in the same way as mosses and ferns. This phenomenon has been observed for different fertilization processes, either when male and female organs are on different seaweeds (such as *Fucus serratus, Ectocarpus siliculosus, Cutleria multifida,* and *Desmarestia aculeata*), or when male and female organs are on the same seaweed (such as *Desmarestia viridis* and *Chorda thormentosa*).

In the first case sexual attraction occurs along different subvariants of the fertilization mode. Thus, swimming spermatozoids released by male organs of *F. serratus* are chemoattracted by non-motile female organs. In contrast, it is the female organs of *E. siliculosus* which chemoattract the isomorphic male organs, while the relatively large swimming female organs of *C. multifida* chemoattract the very small swimming male organs.

In the second case an additional phenomenon is shown by *Chorda tormentosa* (Laminariales) which grows in the northern waters of Norway. Metabolites produced in the eggs of this seaweed, to the amount of one milligram per some hundred million eggs, have the function of triggering the release of spermatozoids before sexual attraction.

MARINE PRIMARY PRODUCERS

The specificity of these chemoattractants, which are small hydrocarbons, spans a vast horizon. Thus, the same compound may act as a sexual attractant for members of taxonomically distant groups or the activity may be restricted to taxonomically close genera, such as *Desmarestia* and *Syringoderma* of north Atlantic waters or *Hormosira, Durvillea, Xilophora, Scytosiphon,* and *Colpomenia* of Australian waters.

The activity of other compounds is restricted to either a single genus of brown seaweed, such as *Ectocarpus, Fucus* or *Desmarestia,* or even to a single species, such as *Dictyota dichotoma, Cutleria multifida* or *Ascophyllum nodosum.*

Apart from these last three cases, the selectivity of these sexual attractants is meager and reflects Nature's economy in not making things more complex than the case requires. In addition, these hydrocarbons are not active in all species of seaweed mentioned above. For instance, a hydrocarbon secreted by the reproductive organs of the brown seaweeds *Spermatochnus paradoxus* and *Ascophyllum nodosum* is only sexually attractive for the latter species. The compound is believed to have no scope with *S. paradoxus* and to be merely the result of a residual chemical mechanism of phylogenetic precursors or just a waste product. Oddly enough, one of the sexual attractants of brown seaweeds, ectocarpene, is also produced by the South African herb *Senecio isatideus.* These might be arguments for Nature's lack of economy, a conclusion that I dislike; it may well be that *S. paradoxus* and ectocarpene have not yet been fully understood.

BROWN SEAWEEDS AND THE OCEAN SMELL

Certain brown seaweeds contain small hydrocarbons which, although structurally similar to the sexual attractants above, act as odoriferous compounds. Notably, a most appreciated mixture of flavors of seaweeds belonging to the genus *Dictyopteris* is responsible for the "ocean smell". Chemists have not missed the opportunity and have carried out the total synthesis of such structurally-simple marine perfumes instead of carrying out the heavy job of collecting tons of seaweeds and wasting a lot of expensive solvents for extractions.

The odor of such hydrocarbons strictly depends on their detailed molecular structure; a high specificity that contrasts with the poor specificity of sexual attractants of these seaweeds.

The selectivity of our nose is not confined to marine perfumes only. Think of the sommelier, who largely relies on his nose in distinguishing wines. The purpose of such a high selectivity in perceiving and discriminating odors is related to the important function of primarily detecting poisonous compounds, from which our survival may depend.

GREEN SEAWEEDS AND COLONIAL SEA SQUIRTS: THE ENDOSYMBIONT THEORY OF EVOLUTION

The photosynthetic apparatus of red seaweeds resembles that of cyanobacteria. Is mere chance? The answer is no, as we believe that cyanobacteria have been incorporated into non-photosynthetic hosts to give origin to the plastids of the red seaweeds. This is an example of the endosymbiont theory of evolution, which is one of the principal theories concerning the evolution of species.

You ask: "So what ? And what now ?" To such well-known question, molecular biology (on the basis of nucleotide sequence of genetic material) has just answered that the symbiosis above occurred relatively late along the protist diversification. This is in agreement of what can be deduced (as discussed in subsequent chapters) from secondary metabolism in a related case of evolution concerning green algae. The precursors of green algae were unknown until 1977, when prokaryotic green algae were discovered to be extracellular symbionts of certain colonial ascidians belonging to the Didemnidae family. These ascidians are small-size sea squirts which are organized in crust-like colonies which secrete calcareous material to form calcareous spicules. When these animals are immersed

53

alive in alcohol, as I used to do, there is a copious evolution of carbon dioxide.

The prokaryotic green alga harbored by some (though not all) Didemnidae is *Prochloron didemni*. This alga contains chlorophyll-a and -b, like eukaryotic green algae and higher plants, but none of the accessory photosynthetic pigments of cyanobacteria and red algae.

FOOD AND POISON FROM GREEN SEAWEEDS

Some green seaweeds are eaten or are used for food products, like the genera *Monostroma*, cultivated in Japan for a product called "aonori", and *Enteromorpha* ("awo-nori"). *Ulva lactuca* (besides being a social problem because of over-reproduction in northern Adriatic Sea) is eaten like a salad even outside Japan and *Prasiola japonica* and various species of *Codium* are also eaten.

Only certain species of *Caulerpa* (which are used in the Philippines for salads) may present health problems. Thus, *Caulerpa racemosa* has a peppery taste and provokes when ingested a mild numbness on the tip of the tongue, a weakening of the legs and in some people breathing difficulties. After being passed around the world, a material (caulerpicin) extracted from *C. racemosa* of the Philippines and reputed to be responsible for the above effects was identified as a mixture of metabolites (formed from a fatty acid and a long-chain amine), although no test of physiological activity was reported. The seaweeds of the genus *Caulerpa* present an interesting case. Natives in the Philippines avoid eating such seaweeds during rainy months, which is probably not just the result of mere superstition; the production of toxic principles is elicited by mechanical

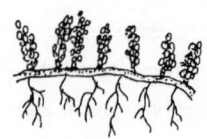

Caulerpa racemosa

injuries in the seaweeds, which may well occur during rain months which are periods of storms and agitation of the waters.

A similar situation occurs in the production of antifungal and antibacterial substances by terrestrial plants subjected to either microbial or mechanical injuries. These compounds are called phytoalexins, a word derived from ancient Greek to mean "warding-off".

Water-soluble polysaccharides containing sulfate groups can also be extracted from green seaweeds but have no market. Seen from a medical point of view there is nothing particular that merits discussion with the green seaweeds either.

MEMORY EFFECT IN HIGHER PLANTS

Higher plants originated from green seaweeds and it should not be surprising that there are natural products which are common to both groups of vegetables. This is a kind of memory effect where the higher plants remember the synthetic ability of their phylogenetic ancestors.

One such example is 'tydemaniatriterpene', a terpene contained in *Tydemania expeditionitis* (Weber-van Bosse), which is a green seaweed from Guam. In fact, this terpene has a structure closely resembling cycloartenol, which is an intermediate in the biosynthesis of sterols in higher terrestrial plants. Another case is presented by friedelin, a terpene that since last century is known to be a constituent of the cork from the European oak and was recently isolated from the green seaweed *Monostroma nitidum*.

DAMSELFISH AS GARDENERS AND OTHER STORIES OF GRAZING ON CORAL REEFS

Seaweeds are not only appreciated by the Japanese and Chinese but by certain marine animals as well. In temperate waters the principal grazers are mollusks and

55

sea urchins, whereas in tropical waters, and especially on coral reefs, the principal grazers, besides sea urchins, are fish. It is also relevant that amphipods (which are small crustaceans) feed on filamentous microalgae.

Herbivorous mollusks have a toothed, chitinous mouth (the radula) which is used to pulverize seaweeds. In place of a mouth, sea urchins have a complex apparatus, called Aristotle's lantern, which is a name that takes me back to my enchantment with the humanistic studies of my youth. This apparatus has the same function and the same specialization for certain algal morphologies as the teeth and beaks of tropical herbivorous fish.

Whereas seaweed grazers exist in all seas, the phenomenon passes unnoticed in the abundance of seaweeds in temperate waters. In the tropics, where seaweed is scarce, grazing by herbivorous species acquires great importance. This is particularly true on the small reefs along the western American coasts. Due to a large tidal range (which approaches six meters in certain areas), the reef is fully covered with water at high tide and thus easily accessible to all herbivorous fish, and as a result seaweeds are rare in such areas.

Even on the large coral reefs along the eastern American coasts, typically in the Caribbean where the reef flat top is always protected by waves, seaweeds are scarce. Sea urchins (mainly of the genus *Diadema*) and herbivorous fish (mainly the parrotfish, such as *Scarus scaber*, but also the surgeonfish) are responsible for algal depopulation.

The imbalance of seaweeds with respect to corals in the tropics lies more in terms of the biomass of the algae than in their variety. For instance, the Great Barrier Reef harbors over

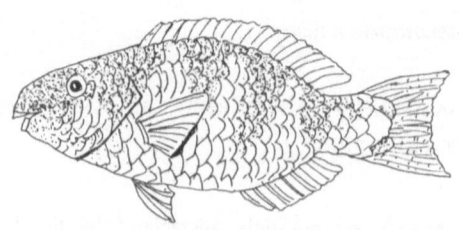

a parrotfish (*Scarus scaber*)

SEAWEEDS

300 different species of seaweeds that colonize the areas left free by the corals. In any event, seaweeds leave space and food to corals, which are the principal builders of coral reefs and are strongly dependent on light for the photosynthetic activity of their microalgal symbionts. These symbionts replace the seaweeds in the photosynthetic supply of nutrients, making the coral reef a rich oasis in the nutrient-poor tropical waters. To avoid being washed off, the microalgae are incorporated into the corals which has the advantage of making the process of food transfer from algae to host particularly efficient.

If grazers of seaweed in tropical areas are kept away by protective nets, overgrowth of seaweed is observed which causes the death of the corals. Following the mass mortality of the sea urchin *Diadema antillarum*, possibly due to a bacterial infection, this is currently a problem in the Caribbean; the presence of this sea urchin has been reduced to 2% of

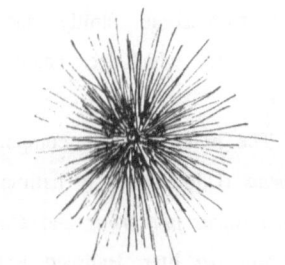

Diadema antillarum

the previous level. As a consequence of less grazing pressure, the algal turf which was a few millimeters thick has now increased to some centimeters and macrophytes have begun to develop. However, either owing to the lack of fecal excretions from sea urchins or to self-shadowing, the seaweed productivity is lower than usual.

Other than with sea urchins, the balance of seaweeds in relation to corals is maintained in Nature by following three strategical lines: by physical or chemical protection of the seaweeds and the intervention of a special gardener. The physical defense of coral-reef seaweeds is a calcareous impregnant in the aragonite form of calcium carbonate that is harder than the calcite form of the skeleton of reef-forming corals, the scleractinians. Some calcareous green seaweeds grow in the exposed windward area of the reef where grazing is uncomfortable to any species. These algae, by cementing calcareous materials together, provide a hard and smooth surface which protects the external reef from erosive forces.

In any event, with respect to the food pressure of the tropics in the internal part of the reef, the calcareous impregnant of seaweeds would not be an adequate deterrent against grazers. This is dramatically shown by certain boring green algae of the genus *Ostreobium* which, though they are able to penetrate the coral mass, are grazed by parrotfish which beat the coral. Therefore, the exposed, calcareous green seaweeds of coral reefs had to start synthesizing compounds that are toxic to herbivores. Such a protection has been developed by calcareous seaweeds belonging to the Udoteaceae family and seaweeds belonging to the genus *Caulerpa* (Caulerpaceae), which lack calcified materials and are therefore particularly exposed to grazers.

The principal toxic compound of various seaweeds belonging to the genus *Halimeda* (Udoteaceae) is halimedatrial, a terpene which can kill the damselfish *Eupomacentrus planifrons* and *Duscyllus aruanus* within one hour at a level of 5 milligrams per liter. Halimedatrial also blocks the sperm development of the sea urchin *Lytechinus pictus*. Halimedatrial exerts a variety of other biological activities, such as inhibiting the growth of marine bacteria and fungi, including species that settle on marine algae of the Caribbean. *Caulerpa bikiniensis* in the Western Caroline Islands and *Udotea flabellum* in the Caribbean contain defensive terpenes as well. The first one contains 'caulerpatriene' and the second one udoteal, which is a structurally similar, though bigger, molecule.

A phytotoxic (to johnsongrass) halogenated terpene from the green seaweed *Neomeris annulata* of Bermuda waters, is also known.

So much for the physical and the chemical defense of green seaweeds. The other strategic line of defense of these algae involves a special gardener, the damselfish. These fish, though they do not feed on corals, eliminate corals by nipping away the polyps; in areas of dead coral the damselfish weed their gardens and in this way control the composition of seaweed species. They also protect their farm quite aggressively from other herbivorous fish, though sometimes they loose the battle against hordes of invaders.

HUMAN INTRUSION AND TRAGEDIES ON CORAL REEFS

Although the grazing pressure in the tropics is very high, in most tropical areas the low level of nutrients does not allow the same development of seaweeds as in temperate waters. Although coral reefs are oases of abundance in nutritionally poor tropical waters, it is only in certain tropical areas, like Martinique, that there is a high level of nutrients and an abundance of seaweeds. In general, overgrowth of algae would threaten the coral reef.

When considering coral reefs one must bear in mind that these ecosystems have been heavily modified. One of the reasons is that natives were used to fishing as a means of meeting their nutritional needs but, following the introduction of our economic system, they have fished more to earn more money since. This has led to an attenuation or to the complete abolishment, as in Hawaii, of the severe traditional fishing regulations which were typical of the tropical, oceanic communities.

Human pollution represents another threat to coral reefs: in Kanehoe Bay, which is on the Island of Ohau, Hawaii, the patch reef was destroyed by sewage.

As a result of this and other modifications in local habits, modern studies of most coral reefs should strictly refer to the present situation. If the causes of disturbance are removed, the coral reef, according to calculations, will take in between ten to a hundred years to return to its previous state.

Human tragedies are also a result of the modification of coral reefs, such as the relocation of Bikinians from their homeland, Bikini Atoll, in the northern Marshall Islands. The tragedy of this self-sufficient population, which lived from fishing and from farming to a lesser extent, began in 1946 when the Americans started to use Bikini Atoll for testing nuclear weapons. The Bikinians were initially relocated to the small Rongerik Atoll, where they did not have enough to survive, then to the US military base of Kwajalein Atoll and finally to Kili Island in the southern Marshall Islands. Islands are different from atolls: in Kili there is neither a lagoon nor sheltered fishing ground, and these differences required that the Bikinians change their way of living.

To be fair, it is not only man but also microbes and animals that may threaten the coral reef ecosystem. Examples that will be dealt with later are the hawksbill turtle (which devours large amounts of sponges), the starfish *Acanthaster planci* (L.) (which destroys corals), and pathogenic bacteria and fungi (which, from time to time, are the cause of mass mortality of sponges). Nevertheless, except for the hawksbill turtle, these other phenomena may well be a consequence of imbalance of coral reefs brought about by human intrusion.

Oceanographic anomalies also have an impact on coral reefs. Hurricanes and typhoons carry out coral colonies, and in the process destroy large parts of the reef. Moreover, there is a sporadical rise in the surface temperature of tropical waters, a phenomenon called "El Niño Current". This was exceptionally strong in 1982-83: in the eastern tropical Pacific the water temperature remained at 4 °C above the normal level for several months, which was the cause for the destruction of large coral reef areas chiefly in the Galápagos Islands, Panama, Tahiti, and the Society Islands. The difference between the changes in coral reefs which are caused by natural phenomena and those resulting from human intrusion is that the former fulfill a balanced plan by Nature concerning destruction/reconstruction, while the latter bring about an imbalance in such plans and are directly responsible for net destruction.

In any event, the common image of coral reef areas as a paradise where living demands little effort is an image created by fiction and movies. Except for fish, there has always been a shortage of proteic food in coral-reef islands and atolls, and the conditions do not allow the intensive fishing that our economic system requires while ciguatera represents a hazard in any case.

ON THE RESEMBLANCE OF SEAGRASSES TO TERRESTRIAL PLANTS

Seagrasses have a climatic zone distribution, although there have been abrupt changes in certain areas recently. A notable case is the regression of *Posidonia* meadows in the Mediterranean, which is of much concern since seagrass meadows represent a particularly important part of the ecosystem and are essential to the life of many marine organisms. Seagrasses are important producers of oxygen, organic nutritious detritus, and also have the major function of holding together loose soil.

Mediterranean and South Australian areas are mainly colonized by *Posidonia* and, to a lesser extent, by manatee grass (*Cymodocea*), whereas Indian and West Pacific areas, as well as the Red Sea, are rich in turtle grass (*Thalassia*), *Enhalus*, *Halodule*, *Syringodium* and *Cymodocea*. North Atlantic, north-eastern and north-western Pacific areas harbor eel grass (*Zostera*) and the Caribbean and surrounding regions turtle grass and *Halophila*, the latter being present in the Red Sea, as well.

Seagrasses resemble terrestrial higher plants in the way in which they extract soil resources through their roots and produce complex polymers, lignins, which serve to strengthen cell walls and make plants woody. However, seagrasses are structurally less rigid, and lack the orifices (stomata) which are typical of higher terrestrial-plants leaves.

Seagrasses are a very scarce source of natural products. We know only of some terpenes from *Amphibolis antartica* of Shark Bay, Western Australia, and some

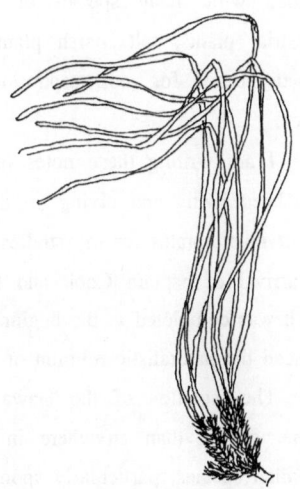

Posidonia oceanica

61

phenolic compounds from *Posidonia oceanica* of the Mediterranean. The phenolic compounds (which are a structural type that is common in higher plants) probably have an antibacterial and antifungal role, just like seaweed phenols.

Given their taxonomical position, not much more could have been expected from seagrasses: their biosynthetic ability is in no way inferior to that of non-marine members of the related families Potamogetonaceae and Hydrocharitaceae.

SALT MARSHES, MANGROVE SWAMPS, AND COCONUT PALMS: A DISENCHANTING VIEW

Emergent coastal plants form productive intertidal shore systems with an important seasonal exchange with the sea of organic material, which is derived from dead leaves and other parts of plants. At mid and high latitudes such a habitat (called salt marsh) generally comprises grasses, reeds, herbaceous plants, and low shrubs, blended with many species of benthic seaweeds and microalgae. Contrary to terrestrial plants, salt marsh plants live in a nearly anaerobic substrate and the oxygen needed for root respiration is obtained through the hollow air tubules of leaves.

I am writing these notes in a cottage on Coconut Island, Hawaii, where I have been living and diving on the coral reef for the last few weeks in order to collect invertebrates for my studies. The Islands of Hawaii were quite isolated up to the arrival of captain Cook and his fleet two centuries ago. It was a revolution which was completed at the beginning of the nineteenth century by missionaries who replaced the naturalistic religion of the Tahitians with their dogmas.

The isolation of the Hawaiian Islands is reflected in a lesser number of marine genera than anywhere in the tropics, and in the presence of exclusive Hawaiian species, particularly sponges. This is also true for terrestrial and intertidal organisms, if imported species are not considered. For example, initially there were no mangroves in Hawaii (they were introduced in 1902) and the fauna now covering

EMERGENT PLANTS

the sea shore is less rich than elsewhere in the tropics.

Just behind the mangroves and often mixed with them, there are coconut palms, as in the front of my cottage. You are probably thinking "what a paradise". This is not always the case, as a stormy wind has been blowing and a heavy rain falling without interruption for a week. It is difficult to keep anything dry and I am engaged in a continuous fight against giant cockroaches. This is a rather familiar situation in this tropical paradise which I remedy by putting all my food, including tea bags, into the freezer and by hanging my clothes from the ceiling.

CHEMICAL SELF-DEFENSE OF MANGROVES

Salt marshes and mangrove swamps are habitats which are quite exposed to human intrusion in terms of dredging, land filling, and the discharge of pollutants. Therefore these habitats, despite their recognized useful assimilative effects (as in the case of *Spartina* salt marshes, which function as natural filters for sewage treatment) are destined to be drastically altered by human intrusion.

Apart from the problems of human intrusion, salt marsh plants and mangroves are naturally subjected to the pressure of many grazers. In response to this the mangroves, like the reef-building green seaweeds, have developed chemical defense mechanisms. These are exploited by natives of the Philippines who use the sap from the roots of the mangrove plant *Heritiera littoralis* as a poison with which to catch fish. The poisonous compound of this sap, called heritol, has now been identified: it is a terpene toxic to *Tilapia nilotica* fingerlings at a level of 20 ppm.

MARINE DECOMPOSERS

In the heterogeneous and complex marine environment, which also receives debris from land, waste recycling is particularly important. The role of decomposers is played by bacteria, fungi and protozoa, which are essential for maintaining a steady balance of the life and conditions in the sea. This occurs either by recycling excrement and debris, thereby providing food for other organisms, or by ingesting other microorganisms, which is the initial route for the transfer of living matter in the sea before the food chain starts.

Contrary to the terrestrial environment, the recycling of organic matter does not occur through putrefaction in the sea. This process is hindered by the high salinity of the sea so that dead organisms form a slime which is dispersed on the bottom of the sea. Certain marine sediments are anomalously rich in inorganic salts and organic carbon; although halophilic marine bacteria exist, such a high content of organic carbon is probably the result of an overall less numerous microbial population and cellular adaptations, which results in a lower metabolic efficiency of microbes at such a high salinity.

The protozoa are a large group of heterotrophic eukaryotes that graze on microalgal, fungal, and bacterial biomass. One recently discovered that they also ingest polymers; a case in point is the choanoflagellate *Codosiga* sp. from the salt marsh estuary of Sapelo Island in Georgia, which is capable of ingesting the polysaccharide dextran. The protozoa traditionally also include heterotrophic dinoflagellates (which we have chosen to place in Chapter 2 instead, together with the phototrophic dinoflagellates). With the exception of the latter, nothing is known about the natural products of protozoa, such as heterotrophic green flagellates, ciliates, and ameboid forms. Therefore, protozoa are not further dealt with in this book except for the peculiar fact that microalgal blooms in subarctic regions of the Pacific Ocean are controlled by planktonic protozoans: unicellular zooplankton grazes on microscopic algae so that there is no spring bloom in these areas.

BACTERIA

4 Marine bacteria

The prokaryotic marine decomposers, the bacteria (unicellular or colonial), are the oldest of all living organisms. Multiplication occurs through the growth and division of cells.

Certain bacteria have features so different from all other bacteria (called eubacteria, which also include cyanobacteria) that one has recently proposed to remove them from the prokaryotes, thus establishing the third primary kingdom, the archaebacteria. These include methanobacteria (which obtain energy by converting hydrogen and carbon dioxide into methane), halobacteria (which grow in water of high salinity), and bacteria which use amines or metals as a source of energy. Sulfobacteria (which utilize sulfur compounds as a source of energy) are also often classified as archaebacteria, though there is still some disagreement.

In this book the archaebacteria are discussed together with the prokaryotes in a separate section; with such complex matters as biological classification one has to be cautious in accepting changes.

Bacteria are ubiquitous organisms, as they have not only successfully managed to penetrate all other organisms, but can live as free species, which also holds true for the sea. So far evidence is only lacking for cometary bacteria.

ARE MARINE BACTERIA MERELY DECOMPOSERS?

Codification is difficult in biology, and marine bacteria under the decomposers are no exception. Their role is, in fact, not limited to waste recycling and some of them are not decomposers at all (such as deep sea bacteria, which are generally in a dormant state with scarce metabolic activity).

65

In contrast, estuarine and coastal bacteria actively metabolize sedimentary matter, producing hydrocarbons previously unknown in relation to living organisms. There are also exceptions on the deep-sea bottom, where bacteria and cyanobacteria deposited with the phytodetritus of euphotic surface waters maintain efficient metabolic activity even at a depth of 4500 meters, as proved under simulated deep-sea conditions.

Equally divergent from typical decomposers are symbiotic bacteria, which are essential to the life of the host organism, and pathogenic bacteria, responsible for a selection in which only the more resistant hosts can survive.

The activity of builders is also implied with phototrophic bacteria, a fine point that is discussed below.

PHOTOSYNTHETIC BACTERIA

Certain bacteria are capable of photosynthesis, although they can shift to non-photochemical modes of nutrition when light fails. Contrary to algae and higher plants, which use water, photosynthetic bacteria use other source materials, such as hydrogen sulfide, in reducing carbon dioxide. Since hydrogen sulfide is a highly poisonous gas for humans, from our point of view this process helps to detoxify the environment.

Purple non-sulfur bacteria show the greatest variation in growing modes, as they are able to grow in the dark in the presence of oxygen, or in the light in the presence or absence of oxygen.

CLASSIFICATION OF MARINE BACTERIA: AN INTRIGUING MATTER

It is difficult to come up with a uniform classification for bacteria, as choice is always biased by the specific research interests of the investigator. Although the only absolute method of classification with bacteria is based on the relative content of nitrogen bases belonging to the nucleotides of the bacterial chromosome, from the point of view of marine ecology a useful classification is: planktonic bacteria, epibacteria, and endobacteria.

Planktonic, free-living bacteria are by far the largest bacterial community in the sea and we now know that they occur at all depths. Books written twenty years ago tell us that seawater at a depth of 500 meters contains only some hundred bacteria per milliliter, while today it is apparent that their number is many times higher.

Until recently undervaluation of the planktonic bacterial population was due to inadequate methods, mainly based on cultures on agar supports wherein planktonic bacteria failed to grow. It was only when direct optical observation and chemical analysis of bacterial compounds became possible that the hidden bacteria could be discovered in huge numbers. The fact that this was only possible through special methods of direct optical observation is no wonder once one realizes that planktonic bacteria are particularly minuscule. Colonies barely attain the size of one thousandth of a millimeter and solitary cells are even smaller. In contrast, epibacteria, which live on other organisms or inanimate matter, are larger and thus easier to study than planktonic bacteria.

Endobacteria are another group of bacteria that live within the cells or in the intracellular matter of other organisms. They can either be the symbiotic type (helping the host) or the pathogenic type (attempting to disable the host).

BACTERIAL CELL WALLS

Another widely-used distinction, according to their reaction to the Gram stain, is gram-positive and gram-negative bacteria. This is an empirical method of treating bacteria with crystal violet dye, then iodine, and finally safranine dye.

The majority of marine bacteria are gram-negative, including the planktonic, free-living forms and most epibacteria. Only in near-shore waters and the nutrient-rich surface-microlayer of seawater do gram-positive bacteria account for a sizable part of the bacterial population.

Despite the fact that gram-positive bacteria are a minority in the sea, a group of them, the Actinomycetes and in particular the genus *Streptomyces*, are of specific interest. Their name recalls the antibiotic streptomycin from terrestrial *Streptomyces*; as a result from its tuberculostatic properties, streptomycin has become an important complement to the penicillins.

The Actinomycetes grow in the presence of air, and some terrestrial species are easily cultured either for antibiotic production, or as biological reactors which are capable of a few but highly specific reactions. These are useful in transforming many substances of pharmaceutical interest, in particular steroids.

One may ask whether the gram-negative cell represents an evolutionary advantage; the answer lies in the organization of the cell wall. Results from observations with an electron microscope and chemical studies depict the gram-positive cell as being relatively simple. The cytoplasm is surrounded by a cytoplasmic membrane and an external thick, rigid envelope of a polysaccharide-peptide polymer called peptidoglycan together with glycerophosphate polymers called teichoic acids. Though simple, such a barrier protects the cell from osmotic lysis and has antigenic activity. Gram-positive bacteria can not survive without cell wall, a circumstance exploited in the treatment of bacterial diseases with penicillins, which inhibit the biosynthesis of peptidoglycans.

The gram-negative cell contains additional, more elaborate cell walls. In the simplest case, the cytoplasmic membrane is surrounded first by a thin layer of peptidoglycan and then by an outer layer of proteins, phospholipids, and lipopolysaccharides, which are polysaccharides bound to fats. Among their other functions, cell membranes act as selective filters for molecules and ions. The multilayered, highly selective, membrane of the gram-negative cell confers to these bacteria an extreme adaptability to the conditions of marine life.

MARINE BACTERIA AS STONE EATERS

With phototrophic bacteria the organizational advantage of the gram-negative cell is coupled with the capacity to carry out photosynthesis by using a variety of substrates, and to shift to a non-photochemical, absorptive mode of nutrition when light fails.

By utilizing inorganic compounds other non-photosynthetic gram-negative bacteria use a different strategy to survive. Thus, nitrifying bacteria oxidize ammonia, which is the end product of decaying proteins, while other bacteria utilize sulfur or other inorganic compounds. We can say that they feed on stones, from which the term chemolithotrophic bacteria arises. Curiously, sulfur-oxidizing filamentous bacteria of the genus *Beggiatoa*, collected by a submersible in the Guaymas Basin vent in the Gulf of California at a depth of 2,000 meters, are unusually large (filaments of up to 0.1 mm in diameter) due to the presence of a large liquid vacuole in the interior of the cell. In such a habitat these bacteria unusually grow essentially as monocultures forming thick mats (3-30 cm).

This complex armor of cell walls and nutritional modes enables gram-negative bacteria to be adaptable to the marine environment, where they play multiple roles. For instance, bacteria which grow on inorganic substrates are involved in important biogeochemical processes in the sea. In their ability to adapt to different chemical and physical conditions, bacteria can only be rivalled by man's adaptability to certain political conditions.

MARINE BACTERIA WITH AN INTERNAL COMPASS

The survival strategies of such simple organisms as bacteria are most varied indeed. Some marine bacteria, when detached from the sediment, swim to the North as if they had an internal compass. That this is really the case can be shown by placing a magnet nearby which reverses the North direction. Under such conditions, these bacteria consistently swim toward the false North. The internal compass of these bacteria is formed by iron-rich cellular particles. What remains obscure, however, is the precise function of such a peculiar survival strategy. The failure to understand this point is our fault; we would be quicker to rationalize the role of an internal compass in a migratory bird.

MARINE BACTERIA AS ARCHITECTS IN MAKING ANTIBIOTICS AND TOXINS

That natural products are involved in bacterial control is shown by the seasonal regulation of the epibacterial settlement on seaweeds. The responsible compounds are phenols secreted by the algae as a protective surface layer. Though algae need bacteria (principally for the production of vitamins) and show aberrant growth in the total absence of bacteria, overproduction can be negative, the brown seaweeds *Sargassum fluitans* and *S. natans* being a dramatic case in point. These seaweeds begin their voyage from the Sargasso Sea in a healthy condition, their tips free of epiphytic bacteria as a result of simple antibacterial phenolic compounds of their cuticle. When these seaweeds grow older, which sadly occurs in all living things, and are no more able to produce phenols, they are invaded by hordes of bacteria and die.

Marine bacteria produce secondary metabolites, too. Gram-negative bacteria, such as *Beneckea gazogenes*, which lives in salt marsh mud, and *Alteromonas rubra*, which is common in autumn in Mediterranean coastal waters, produce bright-red nitrogen compounds of a type already known from gram-positive terrestrial Actinomycetes. These pigments (which are of some concern in the plastic industry as the causative agents of pink coloration in polyvinyl polymers) are important regulatory compounds of marine life as they inhibit the growth of both green flagellates and cyanobacteria. One such pigment, prodigiosin, is common to both *B. gazogenes* and *A. rubra*, as well as to another marine bacterium, *Serratia marcescens*. In addition, *B. gazogenes* produces a related pigment, cycloprodigiosin, whereas *Pseudomonas bromoutilis*, isolated from the turtle grass of Puerto Rico, produces a highly brominated simple alkaloid.

A specially relevant case is that of the fish-pathogenic, gram-negative bacterium *Vibrio anguillarum* which produces an iron-transporting compound, called anguibactin, which is strongly involved in this virulence.

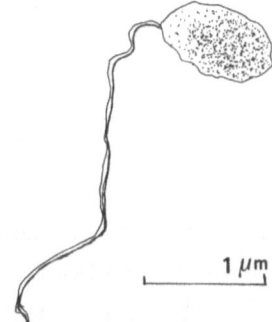

Gram-positive marine bacteria of the genera *Streptomyces* and *Chainia* (order Actinomycetes) produce true antibiotics like their terrestrial congeners. Thus, a marine

Pseudomonas sp.

strain of *Streptomyces griseus* synthesizes ionophore antibiotics structurally similar to those of terrestrial *Streptomyces*. This supports the (controversial) view that marine strains of Actinomycetes, living close to the coast, are mostly adapted species from terrestrial habitats. One of these antibiotics is called aplasmomycin because it inhibits the growth of bacteria and *Plasmodium*. It is only produced when *S. griseus* is cultured on powdered tangle weed, which further testifies that the production of secondary metabolites by microorganisms depends on the nutrition considerably.

Aplasmomycin is an acetogenin with a complex, though highly symmetrical structure that makes its biosynthesis in *S. griseus* simpler; this bacterium only has to make a molecular structural element and couple it in pairs in order to obtain the complete molecule. A similar symmetry-based strategy lies at the basis of the first total synthesis of aplasmomycin.

The aesthetic canons laid down by Nature, the architect, are also manifest in the symmetrical forms of plants, animals, and inanimate objects, such as crystals. That man has always imitated such forms, is evident in the structure of Egyptian pyramids, the Golden Bridge, and the winning boat of the America Cup where the search for symmetry, simplicity, and essence served as guidelines.

There is only a little more to say about the antibiotics of marine Actinomycetes. The antibiotic SS-228Y, produced by *Chainia* sp., is structurally much simpler than aplasmomycin; it is an aromatic compound structurally similar both to the aromatic antibiotics of terrestrial actinomycetes and to xestoquinone of the sponge *Xestospongia sapra*. Although little is known about SS-228Y, terrestrial analogs display antitumoral activity.

Marine bacteria are also known to produce toxins, such as tetrodotoxin (a powerful alkaloidal toxin which is also found in certain fish, starfish, crabs, mollusks - cephalopods, like the blue-ringed octopus, and snails - newts, and frogs) and palytoxin and surugatoxin (found in zoanthids of the genus *Palythoa* - Chapter 7 - and in the mollusc *Babylonia japonica* - Chapter 8 - respectively).

There is now much interest in marine bacterial products, an area where major American and Japanese research institutions have started to cooperate. As a first result, a deep-sea bacterium has given macrolides which *in vitro* inhibit both tumors and the AIDS-causing virus.

THE MILKY SEA REVISITED

Having chosen to place marine bacteria with the decomposers, I did not mention luminous bacteria when considering the milky sea in Chapter 2. Actually, luminous bacteria, most of which live in the sea, contribute much to the luminescence of the sea and are also responsible for the luminescence of certain fish and squid, alive and dead.

Though most luminous fish are deep-sea predators, they can be seen near the surface under special circumstances. This occurs twice a month in the Strait of Messina in the Mediterranean, when turbulence of the water and whirlpools (which are particularly violent if stimulated by strong winds blowing from the south) bring deep-water fish close to the surface. Due to the gravitational attraction of the sun and the moon, the level of the Ionian Sea rises slightly at a certain time. The high tide forces the Ionian Sea toward the Tyrrhenian Sea and (since the Strait of Messina which connects the two seas is a shallow, short channel) a strong stream is the result. A few hours later, the process is reversed: high tide on the Tyrrhenian side pulls the waters toward the Ionian Sea. Twice a month, when the sun and the moon align, their gravitational forces are added and the level of the sea increases beyond normal levels. The stream in the Strait becomes extremely turbulent and whirlpools are formed. In the period between the two tides the waters are still. If you choose the right day when the gravitational forces of the sun and the moon are counterbalanced, you may have calm waters for a short dive. However, one should be accompanied by an experienced local diver; one error and you are in the whirlpools of Charybdis where so many sailors have found their end since Phoenician times.

Some deep-water fish that can be observed in the Strait during the turbulence periods are tiny beautiful monsters such as *Argyropelecus* and *Paralepsis*. Normally they live in the dark in deep waters both sides of the Strait and carry flashing torches, perhaps as a signal to the same species, or to attract the prey.

A few luminous fish permanently inhabit shallow

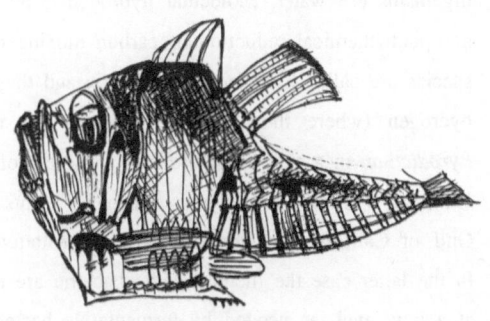

Argyropelecus

waters in coral reef areas and use light to locate prey or communicate. This occurs, for example, in coral reefs of the Gulf of Aqaba, in the Red Sea, and the Comoro Islands, in the western part of

the Indian Ocean with the
flashlight fish (*Photoblepharon*)
which is only about ten
centimeters long and carries just
below the eyes the photophore
organ which can be covered at
will with a fold of skin.

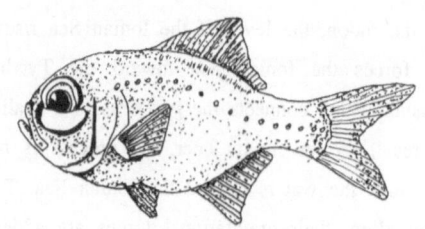

Photoblepharon

The mechanism used by
bacteria to produce light has
been studied in detail:
dihydroflavin monophosphate, a long-chain aldehyde, vitamin-B_{12}, phosphoric acid,
and molecular oxygen are all involved in the luminous processes.

ON MARINE BACTERIA AND METHANE

In photosynthesis certain bacteria utilize molecular hydrogen while more evolved
organisms use water. Molecular hydrogen is also utilized by certain bacteria for the
non-photochemical reduction of carbon dioxide to methane and not to sugars. These
species are called methanogenic bacteria and they live close to geothermal sources of
hydrogen (where they grow at temperatures of up to 110 °C, like the genera
Pyrodictium in a shallow hydrothermal system off Vulcano in the Mediterranean, and
Methanopyrus in the Guaymas Basin hot vents at a depth of 2,000 meters in the
Gulf of California) or together with fermentative bacteria which produce hydrogen.
In the latter case the methanogenic bacteria are responsible for maintaining hydrogen
at a low level, as needed by fermentative bacteria. As expected from this behavior,
methanogenic bacteria when grown in laboratory are strongly inhibited by oxygen,
even in trace amounts.

Methanogenes are not just another group of bacteria; as already mentioned
they belong to the archaebacteria; they lack peptidoglycans in the cell walls, or have

BACTERIA

no cell walls at all, and as a result bear no morphological resemblance to either gram-positive or gram-negative bacteria. The cell walls, when present in archaebacteria, have a special constitution; to this regard a thermophilic methanogenic species of the deep-sea Galápagos hydrothermal vents (called *Methanococcus jannaschii* to remind one of methane production and coccus shape, as well as its discoverer H. W. Jannasch) has been studied in depth. The cell membranes of this species are made up of special lipids, instead of the common fats and steroids (in eukaryotes) and hopanoids (in eubacteria). The special lipids of *M. jannaschii*, besides 2,3-diphosphoglycerate, are terpenes bound to glycerol in such a way that a cyclic, symmetrical molecule is created, like in the above case of aplasmomycin. This is another example of the economy and efficiency of Nature: the high molecular symmetry facilitates the biosynthesis of such lipids and their stability allows *M. jannaschii* to survive the extreme conditions of the Galápagos hydrothermal vents. Help comes also from 2,3-diphosphoglycerate (which is present in enormous amounts in the cell membrane of archaebacteria of the genus *Methanopyrus*) in affording thermostability to the enzymes.

BACTERIA AND FOSSIL SEDIMENTS

Certain bacterial terpenes, called hopanes, were first isolated from recent fossil sediments found in Messel, near Darmstadt in West Germany, where "recent" is related to the geological time scale of sediments, which are 50 million years old in this case.

Hopanoids were later found in all sedimentary rocks in huge amounts. They are also produced by modern bacteria where they have the function of stabilizing the cellular membranes, in much the same way as sterols do with eukaryotes. A typical case is the hopanoid bacteriohopanetetrol. Thus it can be said that the hopanoids are bacterial sterol surrogates. Why Nature has so devised that hopanoids are not degraded and accumulate in such large amounts in sediments remains a mystery, however.

75

LIFE AT THE HIGH PRESSURES OF THE DEEP SEA

The water column is much heavier than the air column so that vertical motions are accompanied by much larger changes of pressure in the sea than above the Earth's surface. For every 10 meters of depth in the sea there is an increase in pressure of one atmosphere. This requires proper instruction in SCUBA diving.

Breathing air at depths below twenty meters for the time required by a marine exploration, or even at smaller depths for longer periods, causes a number of abnormal conditions to occur in our bodies, such as an abnormally high concentration of nitrogen in body fluids. On the way back to the surface, a decrease in pressure allows nitrogen to escape from the body fluids; if this is not done gradually, bubbles of nitrogen may obstruct the blood vessels. As a result, carefully controlled periods of emersion are tabulated for SCUBA diving.

While even normal diving requires a good deal of experience and sixty meters is the safety limit for a skilled diver using compressed air, the use of special mixtures of gases allows for a much wider range. The record was recently set by the French diving company Comex in a successful project of immersion at a depth of 500 meters off the shore of Cassis, in Provence, France. This enterprise necessitated SCUBA diving outside a submersible for long periods of time. Mixtures of helium and oxygen were used up to a depth of about 200 meters, whereas "hydreliox" (which is made of helium, hydrogen, and oxygen in the percentage ratios 49:49:2) was used for greater depths. With so little oxygen, hydreliox lacks the explosive tendency of binary hydrogen/oxygen mixtures.

Many risks in professional diving are now coming to light, however, particularly those concerning divers who service offshore gas and oil plants. The practice of "surface decompression", i.e. emerging rapidly and then entering a decompression chamber, which is economically attractive to commercial companies, carries the risk of damaging the brain, spinal cord, and eyes of the diver. Contrary to previous assumptions, it has now been proved by the medical departments specialized in diving of Australian, British, and American navies, that the time span between when surfacing begins and the full conditions in the decompression chamber, may induce these irreversible damages. Even when all the rules for correct SCUBA diving are followed, a group of doctors of a London hospital believe that damage to the central nervous system is related to frequent diving. According to these physicians, the symptoms begin with the discovery of abnormal changes in the retina, though usually this does not result in visual problems. The obstruction of the blood's flow due to the pressure is believed to be the cause. Whether this is true or not, I am determined to continue to exploring undersea caves using my bottles or simply my lungs.

Deep-sea conditions have not been an obstacle in the adaptation of life. Many organisms, and not only bacteria, live at high pressures. Photographs taken in the Philippine Trench at depths below 9000 meters (where the pressure exceeds 900 atmospheres) show amphipods and other crustaceans in abundance, apart from abyssal cnidarians, brachiopods, mollusks, echinoderms, and fish. The abyssal fauna live largely on phytodetritus which is deposited on the sea bottom from surface waters and contains a variety of organic compounds. Fish is abundant in the deep sea and the fishing industry has developed to the point where it can collect some 32000 tons of the sable fish, *Anoplopoma fimbria*, per year (along the northern border of the Pacific, from California to Japan), 46000 tons of orange roughy, *Hoplostethus atlanticus*, (in New Zealand), 35000 tons of the rough-head grenadier, *Macrourus*, (by the Soviet fleet), and a considerable amount of the scabbard-fish, *Aphanopus carbo*, (from Madeira).

Even more striking is the fact that certain bacteria which live in our garden soil rapidly adapt to the high pressures of the deep sea. This is most surprising in view of laboratory experiments which have shown that high pressures, such as those of the deep sea, are accompanied by various changes in the status of proteins,

77

mainly in the folding of the chains, the association equilibria with ligands, and the nature of reactive groups. This indicates that complete cell suborgans are subject to change under large fluctuations in pressure.

Low temperatures play an important role in the adaptation to deep-sea conditions. With the exception of the Mediterranean basin, where the temperature does not fall below 13 °C between 600 and 4000 meters, and hydrothermal vents, such as in rises of the Galápagos Islands where temperatures as high as 49 °C are normally encountered, typical deep-sea temperatures lie below 4 °C. In such conditions, metabolic processes are slowed down. Though these are conditions which may prolong the life-span, the slow thinking that would accompany such conditions makes this prospect unattractive to us.

The deep sea is much less placid than was thought 15 years ago. At the bottom of the central Indo-Pacific area there are strong earthquakes which, though corresponding to 4 or 5 on the Richter scale, usually pass unnoticed because special waves are generated which are not detected by the standard instrumentation which is in use. Nonetheless, these phenomena have given rise to a V-shaped trench which runs from Guam to western Samoa for more than 5000 kilometers.

The bottom of the deep sea is also disturbed by storms which induce movements of 22 centimeters per second in the waters (as recorded at 5000 meters of depth off the Nova Scotia Rise in northwest Atlantic). Such movements bring about a smoothing of the sea bottom and high turbidity, as has clearly been recorded in photographs.

In deep waters such movements may be relevant to the exchange of sediments with surface waters and must be accounted for in the disposal of radioactive residues. Currently, the deep sea is being used as a sink, a habit that will eventually turn against us.

5 Marine fungi

DISTRIBUTION AND ECOLOGY OF MARINE FUNGI

Fungi are heterotrophic eukaryotes with absorptive nutrition. They are organisms that are quite active on a variety of substrates, the most unusual example of which was perhaps the earliest known manuscript of the Pentateuch, the Aleppo Codex, which was damaged by a fungus of the genus *Aspergillus*.

Fungal cell walls generally consist of chitin, a polysaccharide which carries amino groups and is common in invertebrate animals, too. In some groups of fungi cellulose replaces chitin, or both chitin and cellulose are present.

Marine fungi have the same habitat as a variety of other marine microorganisms and they occur in association with cyanobacteria and green, boring microalgae on calcareous materials, or as parasites of endolithic algae. Though their habitat is varied, marine fungi lack the extreme adaptability and self-sufficiency of bacteria.

Most marine fungi are lower microscopic species; higher, macroscopic species are limited to four genera and do not attain the size of terrestrial higher fungi. Fungi are primarily classified by their different modes of reproduction. Thus, certain fungi reproduce sexually while other ones (mainly yeasts) have a unicellular phase where reproduction occurs by fission or budding.

Although most yeasts need oxygen to grow, some species may facultatively grow without oxygen, e.g. industrially important *Saccharomyces*, which are at the basis of old-established biotechnological processes, such as bread, wine, and beer. This genus is scarcely represented in the sea, however, perhaps as none of these biotechnologies is of interest to marine species.

The role of marine fungi is varied. Certain filamentous species take a highly active part in the decomposition of cellulosic materials of vegetables. Other filamentous epiphytic species serve as attractants for animal predators, thus starting a food chain.

MARINE DECOMPOSERS

A special regulative function is exerted by certain parasitic or pathogenic marine fungi on the growth of fish and crustaceans. Though this is normally a regulative mechanism that serves to eliminate the weakest crustaceans and fish, it may give rise to epidemic diseases in fish and crustaceans that are difficult to control; actually, the major problem in the culture of shrimps stems from fungal infections. We can get an idea of this problem from the price we have to pay for shrimps. Once I asked a Japanese man (an expert in shrimp culture) what they do in case of extensive fungal infections, to which he replied "We just burn the whole plant". Though I am sure that the Japanese dispose of good preventive strategies, this example demonstrates that fungal infections are a major problem in aquaculture.

MARINE FUNGI AND NATURAL PRODUCTS

Lower fungi are omnipresent in the sea; e.g. the surface layer of healthy seaweeds is populated by epiphytic fungi. In contrast, yeasts are scarce as the proteinaceous cuticle of seaweeds periodically renews itself and harbored yeasts are washed off. We also know that a number of regulatory processes in the sea involve fungi.

With regard to natural products, it is well known that terrestrial fungi are the source of important antibiotics such as the penicillins. In contrast, marine fungi have only given a handful of natural products so far, of which some were already known from terrestrial fungi. This is the case of the higher marine fungus *Halocyphina villosa*, which forms tiny fruiting bodies, mixed with a variety of marine

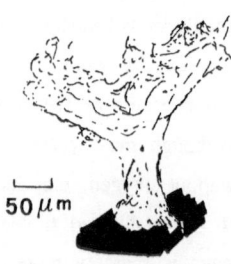

Halocyphina villosa
(sectioned basidiocarp)

invertebrates, on the roots of mangrove trees and produces the same compound (siccayne) as the lower terrestrial fungus *Helminthosporium siccans*. Siccayne is structurally related to prenylquinone, which has been isolated from the colonial ascidian *Aplidium californicum*.

The case of the marine deuteromycete *Asteromyces cruciatus*, which is found on decaying algae in Chesapeake Bay, is similar. In culture, this fungus has given the alkaloid gliovictin, like the lower terrestrial fungus *Helminthosporium victoriae*.

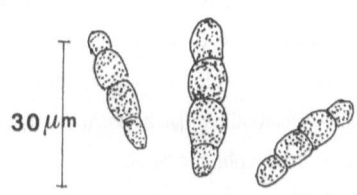

Leptosphaeria oraemaris

(ascospores)

A marine ascomycete, *Helicascus kanaloanus*, which grows among Hawaiian mangroves, is more varied: besides producing the same phenolic lactone, ochracin, as the lower terrestrial fungus *Aspergillus ochraceus*, it also produces unique non-phenolic lactones.

A few natural products are unique to marine fungi. This occurs with the marine ascomycete *Leptosphaeria oraemaris*, which gives the carbohydrate leptosphaerin in culture. Another ascomycete, *Corollospora pulchella*, collected on the beach of Miura City, Japan, in culture produces antibacterial alkaloids, called melinacidins, which derive from tryptophan. Yet another ascomycete, *Zopfiella marina*, separated

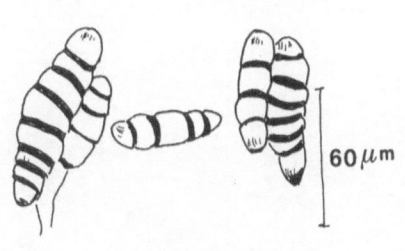

Dendryphiella salina

(conidia)

from sea bottom mud in East China, produces the acetogenin zopfinol in culture.

An example from the deuteromycetes is *Dendryphiella salina* which has been studied in my laboratory, in collaboration with industry, for its ability to grow on the waste of seaweed or seagrass, which is then converted into useful products. If you think of the immense biomass of seaweed and seagrass on certain coasts after a storm at the end of the growing season, you can imagine how valuable *D. salina* might turn out to be. Or one might also introduce the use of *D. salina* in controlling the overgrowth of algae resulting from the eutrophication of coastal waters, such as the northern Adriatic Sea. In any event, cultures of *D. salina* are also a source of new metabolites, such as the terpenes dendryphiellins and a branched carboxylic acid, dendryphiellic acid.

MACROFAUNA

The abundance and variety of macroscopic animals, sessile or moving on the bottom of the sea or in the bulk of the waters, reflect the vastness and diversity of marine habitats. In this immense living system, certain groups of invertebrates (which are animals lacking a supportive skeletal rod - primarily sponges, corals, bryozoans, and mollusks) are at the center of our interest due to the variety of unusual natural products they often contain. In contrast, the chordates (which are animals provided with a supportive skeletal rod during at least one stage of their development) are poor sources of unusual natural products, with the exception of sea squirts and a few fish.

All we know about the natural products of these organisms is limited to shallow water species. Most organisms have been collected manually at low tide or when SCUBA diving at depths not exceeding 60 meters. Small submarines have extended the range to a depth of 350 meters (in exceptional cases, such as in the collection of sulfur bacteria and archaebacteria, 2,000 meters).

Dredging and beam-trawling have allowed for collections up to 600 meters in depth, though this is not the limit of these equipments, as demonstrated in the above examples of industrial deep-sea fishing. Gill nets are rarely employed; I used this device in my expeditions in Brittany waters and took advantage of the methods of the local fishermen who place gill nets in the English Channel. Owing to the violent stream resulting from important tides, gill nets do not only trap fish but also erected sponges in abundance, such as those of the genera *Axinella*

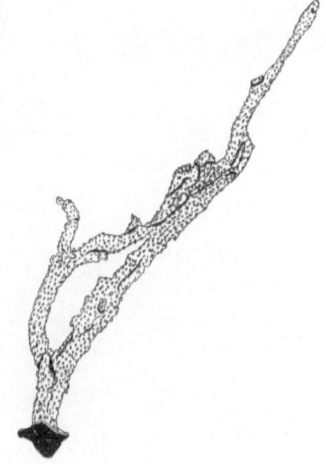

Axinella sp.

and *Raspailia*, which would be difficult or expensive to obtain otherwise.

Our survey below begins just with the sponges, which are the most primitive organisms of the marine macrofauna.

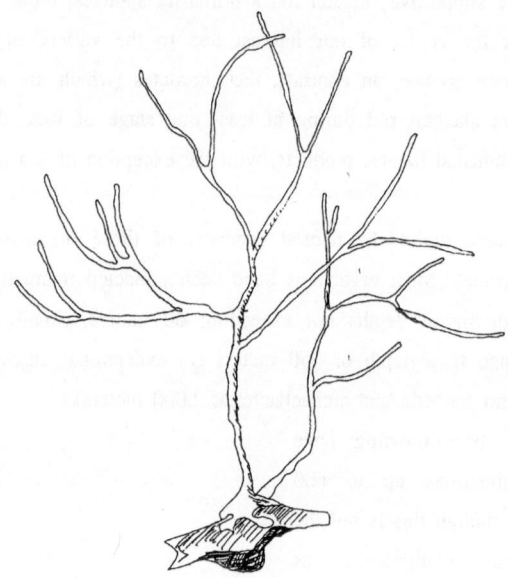

Raspailia ramosa

6 Sponges

WHAT A SPONGE IS

Although my copy of the Webster's Seventh New Collegiate Dictionary dates back to the start of my university studies, I believe that its definition of the sponge as "an elastic, porous mass of interlacing horny fibers...and is able when wetted to absorb water" has not deviated much from the latest edition. It is the blue-blooded image of our bath tool, which is today sadly replaced by disagreeable imitations in pale-yellow plastics with a much poorer ability to absorb water.

I grew up in a sunny temperate area known for its vineyards, my sweet Tuscany, where we bourgeoisely use to measure the performance of sponges with respect to wine rather than water. "He drinks like a sponge", we say in Tuscany of those who drink wine in abundance.

In any event, both definitions of the sponge are biased by the image of the relics of horny sponges hanging in the drugstore and being sold as bath accessories. This image even has an influence on a discipline without frontiers like chemistry, where the sponge used is an expensive "platin sponge" utilized for accelerating the uptake of hydrogen by certain compounds.

To arrive at a less biased image of the sponge, you should either accompany a skilled diver, or explore certain rocky coasts at low tide, such as the coasts of Brittany; most sponges can be recognized by their brightly colored thin crusts or resemble small shrubs. On diving, even the bath sponge appears as a dark mass, which is so different from those relics in the drugstore, that it would go unrecognized by someone who lacks experience.

The sponge is an organism which pumps water in one direction only, taking it from incurrent pores, called ostia, and expelling it from an outcurrent pore, called oscule. This process serves to obtain oxygen and food, as well as for transporting sperm for reproduction and expelling debris.

85

MACROFAUNA

Sponges may be either massive (attaining a diameter of 2 meters in exceptional cases like the loggerhead sponge, *Spheciospongia vesparia*, found in the Caribbean) or such a thin crust that collecting a sufficient amount for the extraction of natural products requires extensive diving.

But perhaps the most intriguing question concerning sponges is: what portion of the biomass specifies them? The answer is that the sponge is neither a colonial organism nor does it possesses organs, although it nevertheless has a tissue organization; it is the whole organism that specifies the sponge.

The sponges are peculiar in that they are able to reorganize their tissues after a disruption. This also constitutes the basis of asexual reproduction which co-occurs with sexual reproduction.

NEGLECT OF SPONGE BIOLOGY AND COMMERCIAL SPONGES

In biology today it is the genetic engineer who is the star. Those in biology who confess that they are not manipulating genetic material, while trying to understand Nature better, are regarded with suspicion. As a result, we are approaching a situation where it will be difficult to find someone who is able to distinguish a seaweed from a seagrass. Regression is rapid; in most marine scientific stations I have problems in having local sponges identified.

Competent sponge taxonomists have become rare indeed; most countries do not have any and things are destined to become worse as economic activities regarding sponges are very limited. Most marine sponges are difficult to sustain in an aquarium and culturing bath sponges would be a most hard venture, probably with little return, because of the commercial success of inexpensive imitations in plastic. Consequently, biology students are more attracted to the culture of fish, mollusks, and crustaceans which offer better job opportunities. Mr. Joubin, Professor at the Muséum des Sciences Naturelles in Paris and the Institut Océanographique in the twenties, failed to be a good prophet when he wrote about the bath sponge

SPONGES

"Aucune autre substance artificielle ou naturelle n'est douée de qualités semblables qui en font la valeur, puisqe rien ne peut la remplacer".

Mr. Joubin was influenced by the importance that commercial sponges still held in his time and by the lack of realization that organic chemistry with H. Staudinger in Zürich between the two world wars was starting to imitate natural polymers. Though I have to admit that the polymers that have been developed so far are not as efficient as the skeleton of the horny sponges (spongin) in absorbing water, they have nonetheless revolutionized our way of life.

In Mr. Joubin's days the sponge market was dominated by the Mediterranean horny sponges. *Spongia officinalis* var. *mollissima*, as the name implies, was the softest of all sponges, and was also called "du Levant" by the high society of Paris. The bourgeoisie could only afford the less expensive *Spongia adriatica*, *arabica*, and *zimocca*. L'"Éponge de Marseille" (*Hippospongia equina* collected in the Gulf of Gabès) was of too coarse a texture to be used for bathing, but the twenties were for well-to-do people.

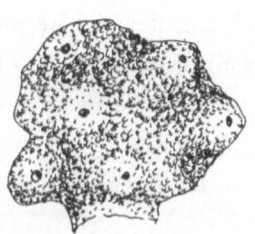

Spongia officinalis

Misleading names were used in Paris too, such as l'"Éponge de Venise". Though I am sure that Casanova's lovers made extensive use of the bath sponge, Venice has always been a shallow lagoon, unsuitable for the queen of the sponges.

Though the air-tight helmet was beginning to replace diving and dredging during Mr. Joubin's times, divers loaded with a stone still collected sponges at a depth up to of 40 meters, which added up to tree minutes diving time. Something nearly unbelievable today, even if the clear waters of the fishing spots in Greece, Turkey, and Syria made diving easier. The sponges were sold in Jaffa, Hydra, Kharki, Tripoli, and Sfax.

Sponges of the Antilles and Central America were also sold in Cuba, the Bahamas, and Florida, but were much less appreciated.

ON EXPERTS, LEONARDO DA VINCI, AND SPONGES

Neglected by the biologists, marine sponges are the organisms preferred most for study by the marine natural product chemists. This is because many sponges are large, easy to collect, and rich in natural products which have no counterpart in terrestrial organisms.

The fact that marine sponges are actively studied today by the natural product chemist, who has a limited or negligible understanding of their biology, could hardly be avoided. It is the price we pay for the rapid development of science, and this development is not illogical. It so happened that the natural product chemist had the methods at hand to open up a new route for the studies of sponges with the discovery of new natural products. Biology students, who are closer to the biological problems of the sponges, could not have been educated *ad hoc* to solve such problems.

Scholars, rather than experts, were the pioneers in this new line of studies on sponges. I am inclined to think that experts are blind to most directions except their own. They perform well in the domain of application of science, while scholars have better chances with the development of scientific principles. However, on the one hand it is also true that the times of Leonardo da Vinci, which were times of profound knowledge ranging from the sciences to the arts, are over because of the tremendous expansion in scientific knowledge, and on the other hand, in some countries, a successful university career is often unrelated to competence and performance. To be in a politically powerful group may be much more relevant. This smoothes out the distinction between experts and scholars, as there are many alleged scholars who lack the competence to solve any problems whatsoever.

SKELETON, REPRODUCTION, AND CLASSIFICATION OF SPONGES

Although the horny sponges have a soft skeleton with no inorganic material, in most sponges the polymeric organic material of the skeleton is reinforced by calcareous or siliceous material. Characteristically, sponges pertaining to the classes Calcarea and Sclerospongiae[6] have a calcareous skeleton, spicular in most of the former and massive in the latter. In contrast, sponges of the classes Hexactinellida and (except for the horny sponges) Demospongiae have siliceous spicules.

Most living sponges belong to the class Demospongiae and are extremely varied in form, habitat, and behavior. From the embryological point of view, originally introduced as a criterion for classification by the French spongiologist Levi, they may be divided into two groups, the viviparous (subclasses Ceractinomorpha and Homoscleromorpha) and the oviparous (subclass Tetractinomorpha). Both viviparous and oviparous reproduction, as well as certain cases of asexual reproduction, involve the formation of larvae, which swim around vigorously until they turn into sessile sponges.

The Calcarea are the simplest sponges, which are generally small and white, or whitish-grey, and live at small depths. In contrast, most hexactinellids live in deep waters and are the least investigated of all sizable animals. It is only in certain areas, such as the subantarctic Indian Ocean, that they can be found in shallow waters.

Hexactinellid sponges are composed of a mass of siliceous spicules with little organic material and are therefore called glass sponges.

The organic skeleton of the sponges is made up of proteinaceous cross-linked polymers, called collagens, which are also present in the bones, teeth, cartilage, tendons, and skin of higher animals. They constitute about 6% of the weight of man and one third of its total protein content. With sponges collagen also takes the special form of spongin.

---

[6]Some specialists do not consider Sclerospongiae to be a class; they maintain that what is now grouped under this name is a polyphyletic group.

89

If boiled, the collagens form a gelatin which is used for food casings and in photographic emulsions; of course waste from common terrestrial animals, rather than sponges, is used for this purpose.

SELF FROM NON-SELF RECOGNITION IN SPONGES

Following the discovery of a natural phenomenon it is in keeping with tradition to give it a proper term; it is a form of amusement which helps the scientist to endure the difficult path of research. Recourse is traditionally made to Latin and ancient Greek, which was also the case with the lectins, the factors responsible for the aggregation of sponge cells.

The name of these low molecular weight proteins, made up of no more than 200 amino acids, is derived from the Latin word *legere* (to select). During the repair of sponge tissues, the lectins are responsible for the recognition of self from non-self with sponge cells by binding themselves to carbohydrate systems. The lectins are not limited to sponges, however; they are present in other invertebrates, in plant seeds and even in certain vertebrates.

All these recognition phenomena involve the interaction of molecules. Active centers of a molecule become bound to receptor centers of another molecule with an extremely high degree of selection that is at the basis of the recognition phenomenon. The fitting of a compound into a receptor follows rules that can be naïvely illustrated with the image of "hand in glove", though this is often an oversimplified view.

PHYLOGENY OF SPONGES AND GEOLOGICAL TIMES

Sponges are very old organisms. Repugnant as it might have been to the archbishop Ussher, some living species are a close copy of those that existed 550 million years ago (the epoch of the first fossil records).

Life of Sizable Organisms on Earth during Various Eras				
Era	Millions of years from the beginning of the epoch	Period (and duration in millions of years)		Common form of life (and first appearance of life)
Cenozoic	70	Quaternary	(1)	(man)
		Tertiary	(69)	flowering plants, insects, mammals
Mesozoic	200	Cretaceous	(60)	reptiles (birds)
		Jurassic	(35)	reptiles, mollusks
		Triassic	(35)	reptiles, (mammals)
Paleozoic	550	Permian	(30)	insects, (reptiles)
		Pennsylvanian	(20)	
		Mississippian	(30)	feather stars
		Devonian	(45)	(amphibian)
		Silurian	(35)	(fish with jaws)
		Ordovician	(90)	fish without jaws, corals, (terrestrial plants)
		Cambrian	(100)	all invertebrates, no chordates [beginning of fossil records]
Pre-Cambrian	3,000			(marine invertebrates)

As to the life history of today's classes of sponges, first the Calcarea and the Hexactinellida and later, starting from the mid-Cambrian era, the Demospongiae appeared. Viviparous species of demosponges represent the latest stage in this evolution while it is thought that the Sclerospongiae diverged from the Demospongiae at some point during the early Paleozoic era.

GROWTH, HEALTH, AND THE ANTIMICROBIAL NATURAL PRODUCTS OF THE DEMOSPONGES

Sponges grow slowly in quiet conditions, with little specific or casual predation, excepting a few fish and the hawksbill turtle in the Caribbean. Boring sponges of the genus *Cliona* grow to be older than twenty years and large hexactinellids are estimated at having been born long before chemistry started.

In general, bacteria are not harmful to sponges and many sponges live in symbiosis with and feed on bacteria via phagocytosis. It is true that there are exceptions: though no true epidemic bacterial diseases of sponges are known to us, sporadic cases of sponges dying due to the invasion of pathogenic bacteria have been recorded. For instance, mass mortality of the horny commercial sponge *Hippospongia communis* is taking place in eastern Mediterranean, which is perhaps due to pathogenic fungi or bacteria. But we can affirm that, in general, sponges are healthy organisms.

The claim that marine bacteria are generally harmless to sponges is also suggested by the fact that sponges in temperate zones show a higher percentage of antibacterial activity than those in tropical zones. Should tropical species have needed protection from bacteria, then they would have developed an antibacterial defense more efficiently than species in temperate waters, where bacteria are less aggressive.

In order to answer the question whether the healthy condition of sponges is the result of a particularly efficient immune response, let me spend a few moments on the immune response of vertebrates. With the tragedy of AIDS, where the immune response is weakened or destroyed entirely, the term immune response has been taken up by the mass media. With vertebrates it deals with a process of recognition in which the role of self from non-self distinction is played by proteins in the body fluids called immunoglobulins. These proteins act as antibodies by attaching themselves to specific places on foreign substances (called antigens) which are thus inactivated. Pathogenic microorganisms or viruses may thus be controlled. Immunoglobulins can be induced by injecting components of the pathogenic microorganism or virus into the body fluids of vertebrates in tiny amounts.

Sponges, like all other invertebrates, lack true inducible immunoglobulins; however, they possess hemocytes with phagocytic activity which, by exerting self from non-self recognition, manage to keep the body fluids free from pathogenic microorganisms and viruses. In addition, marine sponges have developed a system of defense that is based on natural products. This is the result of complex plans: the slow growth of sponges is accompanied by a slow rate of production of secondary metabolites.

The defensive natural products of sponges belong to various chemical classes with structural features that differ considerably from the natural products of terrestrial organisms. Whether sponge cells are responsible for the whole of this diversified production is uncertain, however. A case in point is the polyether okadaic acid: though first isolated from a sponge, it is produced by a dinoflagellate.

Global organic extracts from sponges inhibit marine bacteria (especially of the gram-negative type) more than non-marine bacteria. Whether this reflects the defensive role of the antibacterial compounds in the sponges is debatable. An alternative suggestion has been made that the sponge products inhibit bacteria which then adhere to one another to form particles large enough to be retained as food by the filtering apparatus of the sponges.

A special place in this context is occupied by *Spongia zimocca* which, as mentioned before on passing, is a commercial sponge which was thought to live

93

exclusively in the east Mediterranean and Tunisia. Quite surprisingly, I found that this sponge heavily colonizes a small area in the western Mediterranean, south of Livorno. But even more surprisingly we found that this sponges accumulates halogenated defensive terpenes of close-living red algae of the genus *Laurencia* and probably also elaborates them. No other case is known of uptake of macrophyte metabolites by sponges.

Other studies carried out on the antimicrobial products of sponges have no relevance on sponge ecology, as inhibition tests have been carried out *in vitro* with gram-positive bacteria and fungi of terrestrial origin. These studies warrant presentation here for two reasons, however. The first reason is that the molecular structures of bioactive products have been elucidated in detail, so that there is a lot to learn about the natural products of sponges. The second reason is that metabolites that actively combat terrestrial microbes might prove to be of medical or agricultural relevance.

Beginning with the demosponges of the subclass Tetractinomorpha, *Ptilocaulis* aff. *P. spiculifer* (a rope-like orange sponge of the order Axinellida which lives at easily accessible depths for SCUBA diving along the coasts of the Honduras) contains the alkaloid ptilocaulin which strongly inhibits gram-positive human pathogenic bacteria belonging to the genus *Streptococcus*.

Moving to the southern Pacific we meet other sponges of medical interest. One is *Jaspis* sp. (order Astrophorida) collected at Suva Harbor, Fiji, and in a marine lake of Palau Island in the Caroline Islands. It contains a macrocyclic modified peptide, called jaspamide, which belongs to the same class as didemnin-B, which was isolated from the ascidian *Trididemnum solidum*. Jaspamide inhibits a common pathogenic yeast, *Candida albicans*.

Another finger-like orange Astrophorida, possibly belonging to the family Jaspidae, contains acetogenins which inhibit *Streptococcus pyrogenes*, a bacterium pathogenic to man. Such substances are called bengamides after the place where the sponge was collected, Benga Lagoon in the Fiji Islands.

SPONGES

Turning to the class Ceractinomorpha, a Mediterranean sponge of the order Verongida, *Aplysina* [= *Verongia*] *cavernicola* (Vacelet), which, as the name implies, inhabits caves or shaded areas, produces the antibacterial compounds 7-bromocavernicolenone, bromochloroveron-giaquinol, and the structurally related cavernicolins, which are biosynthesized from the protein amino acid tyrosine. Such products are accumulated by the sponge predator, the nudibranch *Tylodina perversa*.

Both the Mediterranean *Aplysina cavernicola* and *Aplysina fistularis* [= *Verongia thiona*], collected near La Jolla,

Verongia cavernicola and (right, enlarged) *Tylodina perversa*

California, contain other antibacterial tyrosine metabolites, called aerothionin and homoaerothionin, which are structurally related to bromochloroverongiaquinol and were the subject of a unique study. By applying a special technique of X-ray diffraction microanalysis to *A. fistularis*, aerothionin and homoaerothionin were localized within the spherules of spherulous cells in a special cell layer (the mesophyl) just below the external cell layer (the choanoderm) of the sponge. These spherulous cells degenerate through the mesophyl, thus releasing aerothionin and homoaerothionin.

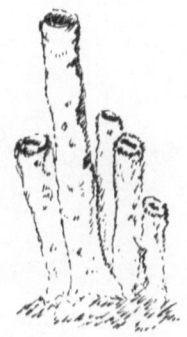

It is in the Caribbean, however, that most antibacterial sponges have been collected. One example is an undetermined pink-colored sponge of the family Plakinidae (Homoscleromorpha) which grows like a shelf fungus on corals in exposed areas at a depth of

Aplysina fistularis

60 m and contains acetogenins, called plakinic acids, which inhibit the growth of the soil fungus *Penicillium atrovenetum* as well as that of the bakers' yeast, *Saccharomyces cerevisiae*.

The reason why we know of so many antimicrobial compounds from Caribbean sponges is that oceanographic expeditions with the vessel R/V Alpha Helix were sponsored in that area at a time, 1978, when antimicrobial activities were still of considerable interest to the pharmaceutical industry.

PREDATION AND DEFENSIVE NATURAL PRODUCTS WITH THE DEMOSPONGES

Sponges have few predators, which go by largely unnoticed. Apart from a few teleost fish, the main predator is the large hawksbill turtle (*Eretmochelys imbricata*). In the Caribbean, in its limited home range, the turtle's powerful jaws are responsible for depleting large amounts of sponges, including the poisonous ones. The process is completed by voracious fish that live off the exposed tissues of the sponge.

Because of this, mass fatalities in coral reef populations which eat marine turtles have been attributed to toxic sponge products retained by the hawksbill turtle. Though this is likely, there is no proof, however.

Outside the home range of the hawksbill turtle, which is very limited, the sponges live in heavens. In the food pressure of coral reefs it is not unusual to observe intact fleshy

hawksbill turtle
(*Eretmochelys imbricata*)

sponges, like *Zygomycale parishi*, which looks like a snaky-haired mythological gorgon, in exposed areas of Hawaii while all around voracious fish are searching food.

Even horny sponges (Dictyoceratida and Dendroceratida), which lack the physical protection of spicules, are not eaten; this is due to antifeedant or toxic compounds present in the sponges, which can be illustrated with various species of *Dysidea* (Dictyoceratida). Thus, *Dysidea herbacea* of Bowl Reef, Eastern Australia, contains a modified sterol, herbasterol, which is highly toxic against reef fish.

Another species, *Dysidea fragilis*, from the coral reef of Coconut Island in Kanehoe Bay of Oahu, north of Honolulu, contains terpenes with antifeedant properties as a preventative against local fish of the genus *Chaetodon*. These terpenes have been called nakafurans from the Hawaiian word "naka" meaning sea creature.

Dysidea fragilis

Dysidea fragilis is a cosmopolitan species which inhabits the northern coasts of France, too. There are no coral reefs in Brittany, but there are cold waters and important tides, such as at the Pointe de Pen-Lan, on the Baie de Morlaix, where I collected *D. fragilis*. None of the nakafurans belonging to the Hawaiian *D. fragilis* were present in our sponge, which was rich in an unrelated terpene instead; with in mind the old Breton name Pen-Lan, the terpene was called penlanfuran. A third collection of *D. fragilis*, this time from the Fiji Islands, was found to be different again. This sponge does not contain terpenes at all but a cytotoxic alkaloid called dysiazirine.

With the last of these sponges we have reached the southwest Pacific. Moving further to the west, to the waters of Laing Island (in the territory of Papua-New Guinea) which is an area infested by predators, we meet the sponges

Carteriospongia [= *Phyllospongia*] *foliascens* and *Fasciospongia cavernosa*, which are taxonomically close to the bath sponge. The former contains a terpene structurally related to the terpene desacetylscalaradial derived from the sponge *Cacospongia scalaris* of Wakayama. The latter contains the degraded terpene cavernosine. Both terpenes proved ichtyotoxic to the freshwater fish *Lebistes reticulatus*. If you are surprised that a freshwater fish, rather than a coral reef fish, has been chosen for this ichtyotoxicity test, you have to consider that structural studies of marine natural products are often carried out far from coral reef areas (this experiment was carried out in Bruxelles).

While ichtyotoxicity is a main characteristic of horny sponges inhabiting coral reefs, a taxonomically distant sponge, the segmented red-colored *Latrunculia magnifica* (Tetractinomorpha, Hadromerida) from exposed zones in the coral reef of Eilat, is exceptional in exhibiting the strongest ichtyotoxicity encountered in marine sponges so far. The responsible toxins, latrunculin-A and the structurally similar latrunculin-B, are acetogenins with the unusual property of disrupting microfilament cellular organization.

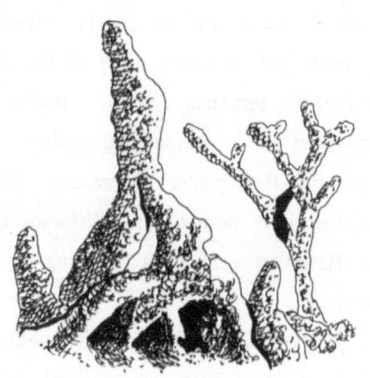

Fasciospongia sp.

Sponges are protected against other predators, too. Thus, *Spongia idia*, a fleshy sponge which inhabits the waters off Pt. Loma (San Diego, California) at a depth of 15 meters, contains the terpene idiadione, which is toxic not only to such a voracious predator as the sea star *Pisaster giganteus*, but also to the brine shrimp *Artemia* sp. and the bryozoan *Membranipora membranacea*.

In this line we can also consider the inhibition of cell division in the fertilized eggs of the starfish *Asterina pectinifera*; the responsible compounds are terpenes contained in the sponge *Cacospongia scalaris* (Dictyoceratida) which grows in the Izu Peninsula waters of Japan. In view of the fleshy constitution of this sponge, which lacks spicules, and the predatory activity of starfish, this might well be a protective mechanism developed by the sponge. Similar activity is exerted by the nucleoside mycalisine-A of the sponge *Mycale* sp. (Poecilosclerida) which inhabits the Gulf of Sagami in Japan.

Among the predators of sponges we have not yet mentioned the most common ones, which are mollusks of the order Nudibranchia. They obtain from sponges both food and antifeedant or toxic metabolites which keep predators away. But this is another story that is presented in chapter 8.

Certain sponges do more than merely defend, instead they attack by burrowing. Whereas most such sponges burrow into rocks, shells, and dead coral, several varieties of *Siphonodictyon coralliphagum* (Haplosclerida) from the Caribbean and the central Pacific burrow into living corals. These sponges also secrete terpenes which inhibit the growth of reef-building corals.

NATURAL PRODUCTS OF PHARMACOLOGICAL INTEREST FROM DEMOSPONGES

While some of the antimicrobial metabolites isolated from sponges may be of pharmacological interest, cellular, antiviral, enzymatic, vascular, muscular, and anthelmintic activities of sponge metabolites are far more appealing today.

Antileukemic compounds have been isolated from a number of demosponges belonging to all subclasses. Let us begin with the Tetractinomorpha. The following examples are from the Caribbean; *Ptilocaulis* aff. *P. spiculifer* (Axinellida) contains the alkaloid ptilocaulin, which we already know as an antimicrobial agent, while *Dragmacidon* sp. (Axinellida), collected at a depth of 148 meters at Sweetings Cay,

MACROFAUNA

Bahamas, contains a tryptophan-derived alkaloid, dragmacidin.

Moving to very shallow waters of the Gulf of California, we find the sponge *Geodia mesotriena* (Astrophorida); it contains antileukemic proteins called geodiastatins after the Greek word *stasis* meaning inhibition.

The sponge *Latrunculia brevis* (Ridley and Dendy) (Hadromerida) was collected by dredging at a depth of 110-145 m off Otago Peninsula, New Zealand. It contains antileukemic alkaloids called discorhabdins from the discorhabd microscleres typical to sponges from the family Latrunculidae. One of such alkaloids, discorhabdin-A, is identical to prianosin-A isolated from the sponge *Prianos melanos* (Halichondrida) found in very shallow waters (2-3 m) off Motobu Peninsula, Okinawa. The batzellines are structurally related alkaloids which have been isolated from the sponge *Batzella* sp. (Poecilosclerida) dredged at a depth of 120 m in the Bahamas.

All other examples of antileukemic sponges below come from Japan. Thus, one *Theonella* sp. (Lithistida) contains the acetogenin misakinolide-A, named after Maeda-misaki, Okinawa, where the sponge was first collected. Misakinolide-A is identical to bistheonellide-A, which was isolated from a morphologically different *Theonella* sp. from the Hachijo-jima waters in the Izu Archipelago. The compound is a macrolide, i.e. a compound from the class of sphinxolide, which was isolated from an unidentified nudibranch collected in Hawaii. *Discodermia calyx* (Spirophorida), found in very shallow waters of the Gulf of Sagami, contains a compound, called calyculin-A, which has mixed biogenesis (the molecule is composed of acetogenin and amino acid fragments).

Theonella sp.
of Maeda-misaki

Antileukemic products are also found in Ceractinomorpha. Thus, *Prianos melanos* (Halichondrida) from Okinawan waters contains a cytotoxic alkaloid, called prianosin-A, which we know to be identical to discorhabdin-A of *Latrunculia brevis*

(Ridley) found in New Zealand waters. *Cribrochalina dura* (Haplosclerida, Haliclonidae) collected at Staniel Cap in the Bahamas contains the acetogenin duryne, which is toxic to colon, lung, and mammary human tumor cells. The *Haliclona* sp. (Haplosclerida) found off Manzamo Island, Okinawa, contains complex alkaloids called manzamines. The sponge *Cacospongia scalaris* of Wakayama waters contains the terpene desacetylscalaradial. Notably, this

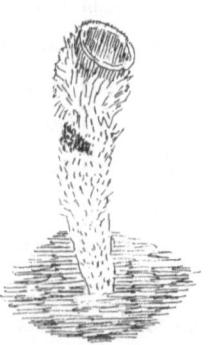

Cribrochalina sp.

terpene is absent from the same species which we collected (by SCUBA diving at 60 meters depth) at Cap de Nice on the Côte d'Azur. *Halichondria* [= *Reniera*] *okadai* (Kadota) contains the acetogenin okadaic acid, which we have already encountered as a product of the dinoflagellates.

The Mediterranean sponge *Aplysina* [= *Verongia*] *cavernicola* (Vacelet) (Verongida) also contains various cytotoxic metabolites of tyrosine origin which we have encountered above as antibacterial compounds.

The sponge *Hyattella* sp. (Dictyoceratida), collected off Manado,

Halichondria sp.

northern Sulawesi, Indonesia, at a depth of 20 meters, contains a macrolidic acetogenin, laulimalide, which belongs to the same compound class as aplysiatoxin (isolated from the cyanobacterium *Lyngbya majuscula*). This compound, which is strongly cytotoxic to KB cell lines, takes its

101

name from the Hawaiian word "laulima" for "people working together" to express the idea of cooperation among different research groups. As a final example, the same type of cytotoxicity is shown by swinholide-A, a macrolide (i.e. a compound belonging to the same class as sphinxolide) isolated from the sponge *Theonella swinhoei* (Lithistida) from both the Red Sea and Okinawa.

Antiviral activity *in vivo* is exerted by mycalamide-A, an acetogenin of the sponge *Mycale* sp. (Poecilosclerida) found in New Zealand waters. The compound is structurally similar to okadaic acid found in the dinoflagellate *Prorocentrum lima* and the sponges *Halichondria okadai* and *Halichondria melanodocia*.

The inhibition of cell division in fertilized sea urchin eggs is shown by various substances produced by sponges of the subclass Ceractinomorpha. A notable case is the terpene-hydroquinone avarol, which was isolated, together with the corresponding terpene-quinone avarone, from the Mediterranean sponge *Dysidea avara*. Avarol is in fact active against AIDS. The other examples are from Japanese waters, e.g. a duryne-like acetogenin isolated from the sponge *Tetrosia* sp. (Haplosclerida) found in the waters off Hachijio-jima Island and the terpene okinonellin-A of *Spongionella* sp. (Dictyoceratida) found in Okinoshima Island waters.

Screenings for enzyme inhibition have mainly concerned the sodium-potassium-dependent ATPase (Na^+,K^+-ATPase). This enzyme is responsible for phosphorylation of adenosine. As examples from the subclass Tetractinomorpha I have selected three sponges from Japanese waters. 1) the *Tetrosia* sp. from Hachijo-jima Island, already discussed above, where the active product is a cyclic peptide consisting of thirteen amino acids; 2) *Agelas* sp. (Axinellida) from Okinawa which contains as active products nitrogenated terpenes called agelasines; 3) the sponge *Epipolasis* sp. of the order Choristida from the Izu Archipelago where the active products are terpenes.

In Ceractinomorpha, acetogenins isolated from the sponge *Siphonochalina truncata* (Haplosclerida) from the Gulf of Suruga in Japan and amino acid metabolites of *Psammaplysilla purea* (Verongida) from Ishigaki Island, Okinawa, inhibit Na^+,K^+-ATPase. With the latter sponge the responsible metabolite, purealin, biogenetically derived from the amino acid tyrosine, exhibits not only $Na+,K^+$-ATPase and myosin Ca^{++}-ATPase inhibition, but also enzyme activation, specifically on myosin K^+,EDTA-ATPase. Similar enzyme inhibitions have been found for other tyrosine metabolites isolated from the Mediterranean sponge *Aplysina* [= *Verongia*] *cavernicola* (Vacelet).

Enzymatic activity has been reported also for the terpene dysideapalaunic acid isolated from a *Dysidea* sp. of Palau. The terpene inhibits aldol reductase, an enzyme involved in the primary transformation of sugars. If dysideapalaunic acid proves to be non-toxic, it might be of interest to diabetics, as a way of preventing complications.

All compounds with vascular and muscular activity have been isolated from sponges collected in Japanese waters. Nitrogen-containing terpenes with antispasmodic activity, called agelasidines, have been obtained from the sponge *Agelas* sp., mentioned above in connection with the agelasines. The active compounds in the sponge *Aaptos aaptos* (Hadromerida) from Okinawan waters and in an unidentified *Agelas* sp. of Kerama Retto, Okinawa, are the alkaloids aaptamine and keramadine, respectively. These alkaloids show antagonistic activity toward serotonergic receptors in the rabbit aorta and α-adrenoceptor blocking activity. The same activity is shown also by the alkaloid hymenin from the sponge *Hymeniacidon* sp. (Halichondrida) of Okinawa.

The sponges *Cryptotethya crypta* and *Dasychalina cyathina* contain spongosine, which is an asystolic nucleoside that also acts on the central nervous system. The nucleoside is structurally related to mycalisine-A discussed above. Examples of compounds with vascular activity are 1) the antispasmodic terpene okinonellin-A present in the sponge *Hippospongia* sp. (Dictyoceratida) which lives in the waters off Keram Island, Okinawa; 2) the cardiotonic aromatic metabolite xestoquinone present in the sponge *Xestospongia sapra* (Petrosiida) which lives in Okinawan waters; 3) the vasodilatative alkaloid xestospongin-C isolated from the

MACROFAUNA

sponge *Xestospongia exigua* which inhabits Australian waters; and 4) a terpene which inhibits platelet aggregation and which was isolated from the sponge *Halichondria* sp. (Halichondrida) collected around Palau Island. This terpene is structurally similar to desacetylscalaradial.

Anthelminthic properties have been found in the nitrogen compounds of an undescribed finger-like orange sponge belonging to the order Astrophorida; the names bengamides and bengazoles were taken from Benga Lagoon, in the Fiji Islands, where the sponge was collected.

Acetogenins of the sponge *Xestospongia* sp. (Petrosiida), the melines, are active against the intestinal parasite *Giardia*. The compounds have structures closely related to duryne, an acetogenin found in the sponge *Cribrochalina dura*.

The sponge *Luffariella variabilis* (Polejaeff) (Dictyoceratida), collected around Palau Island, is as biochemically variable as its name implies. Certain samples of this sponge were found to contain the terpene manoalide while other ones, collected in the same area and morphologically identical, proved to contain luffariellin-A and -B which have the same basic skeleton, but are different in structural details. The name manoalide stems from Manoa Valley where the University of Honolulu's main campus is located and where initial work on *L. variabilis* was carried out. All these terpenes, although they show strong antiinflammatory and analgesic activities, are not structurally related to well established analgesic agents, natural or synthetic, such as morphine, endorphin, salicylates, and indomethacin which are not terpenes. The activities of manoalide and the luffariellins originate from interference with the lypoxygenation of arachidonic acid, with the suppression of prostaglandin and leukotriene formation; remarkably, this is the mechanism which is at the basis of modern synthetic analgesic agents.

Antiinflammatory properties are also shown by the terpene foliaspongin, isolated from the sponge *Phyllospongia foliascens* (Dictyoceratida) found in Okinawan waters.

Most of the "modern" bioactive sponge compounds in this section come from samples collected in Japanese waters. This can hardly be a specific attribute of the sponges which grow in that Pacific area; it is more likely the result of the Japanese's investment and alertness in scientific research. While the situation may be changing now with the involvement of American, French, and Spanish private companies, until recently only Japanese companies had been truly active in the area of bioactive marine natural products, except for a brief period in which Hoffman La Roche scientists worked along the Great Barrier Reef.

Before closing this chapter on demosponges I want to present a *Dysidea* sp. (Dictyoceratida). We collected our sample off the lagoon of Venice where the bottom of the sea is sandy, which makes diving a boring activity. However, if you do have the good fortune to find a few stones, then all benthic species of the sea of Venice are at your disposal. We were lucky to find in such a place a sponge with the morphological characteristics of the genus *Dysidea*. The sponge proved to contain a terpene called adriadysiolide after the sponge and Adria, the ancient name of that area. Although adriadysiolide was easily recognized as being the smallest terpene ever isolated from a sponge, its molecular structure proved difficult to elucidate in detail because of the small amount of the compound at hand. The sponge could not be better identified and, unfortunately, it was never found in that sparse area again. We substituted Nature by synthesizing adriadysiolide, thus elucidating its molecular structure.

HEXACTINELLIDA, CALCAREA, AND SCLEROSPONGIAE: THE LIMITED SCENERY IN SPONGE NATURAL PRODUCTS

Only few natural products are known in sponges of the class Calcarea, and all of them are alkaloids. Although many bioactive alkaloids are known, those derived from the Calcarea are not bioactive. This may be due to the fact that these sponges are adequately protected by a calcareous skeleton, and their small organic content does not attract predators.

MACROFAUNA

Leucetta microraphis from shallow-water caves and under rock overhangs in Harrington Sound, Bermuda, contains the alkaloid leucettidine while *Leucetta chagosensis* (Dendy) of Na'ama in the Gulf of Eilat contains the alkaloid naamidine-A.

I have already commented on our scarce knowledge of the biology of the hexactinellids; we know just as little about the natural products of these sponges. Thanks to the French CNRS oceanographic organization we were among the first to collect hexactinellid sponges from the pre-Antarctic area around the French Crozet Islands. "Glass sponges" seemed to me an appropriate term for these sponges; their spicules were long and as solid as keen-edged small swords. The hexactinellids yielded much glass and only miserable amounts of organic extract. Even worse, the extracts did not contain the variety of natural products found in certain demosponges, although they showed strong herbicidal properties.

In regard to steroids and fatty acids, the hexactinellids that have been specifically investigated so far do not differ much from the demosponges. Our hexactinellids proved to contain a sterone which has already been isolated from demosponges of the order Astrophorida, for instance the Mediterranean *Geodia cydonium*[7]; this sterone has the same skeleton as pavoninin-1 isolated from the peacock sole.

Related observations have been made by scientists from New Zealand; hexactinellids, collected in their home waters, proved to contain the typical fatty acids of the demosponges. All these observations contradict the recent suggestion, made on biological grounds, that the hexactinellids be separated from the phylum Porifera.

---

[7]In demosponges, steroids of this class are formed in a sequence that involves *de novo* synthesis by the sponge's microalgal symbionts, followed by structural modifications carried out by the sponge's enzymes. This requires a strict balance in Nature: constant steroid composition in the sponges is maintained through an obligate symbiosis between the sponge and its microalgal symbionts.

106

SPONGES

Finally, we come to sponges of the Sclerospongiae class. Regrettably nothing is yet known about their natural products and the area is waiting for a patient diver willing to carry out the heavy job of collecting sponges of this class in sufficient amount to allow the isolation and structural studies of metabolites.

7 Cnidarians

To my two European readers, coral connotes jewelry made of the red calcareous coral (*Corallium rubrum*) found in Mediterranean waters or of its less appreciated white or pink type. In the east, coral connotes jewelry made with black coral, the skeleton of the antipatharians. In neither case can the original skeleton be recognized; before they can be used for jewelry, the surface of calcareous corals must be scoured to remove all orifices which harbor the polyps.

Corals were first recognized as animals by Peysonnel, a physician in Marseille in the early eighteenth century. Since then more corals have been discovered, all of which look very different from either *Corallium rubrum* or the antipatharians. Thus, to those who know the tropics, coral means reef-forming corals. These cnidarians are capable of fixing calcium ions from seawater to build up hard calcareous structures which are largely responsible for the formation of reefs and islands in tropical and subtropical waters. These corals belong to the order Scleractinia of a subclass called Hexacorallia as their tentacles and gastric septa number six or multiples of six. These reef builders reproduce via an ovoid ciliated larva called planula, which develops directly into the polyp.

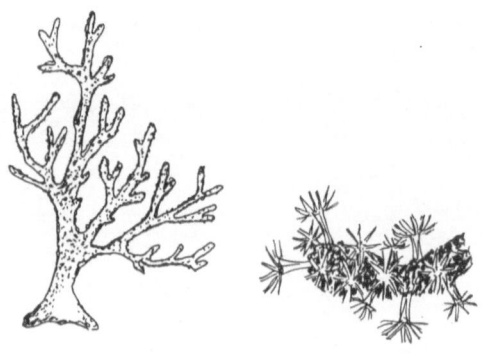

Corallium rubrum

Red, pink, and white precious corals belong to the subclass Octocorallia, order Gorgonacea: their tentacles and gastric septa number eight or multiple of eight.

There are also soft corals, such as *Sarcophyton trocheliophorum*, that do not contribute to the solidity of the reef. In temperate areas, too, most octocorals are soft; a widespread genus is *Alcyonium*.

The phylum Cnidaria also includes the hydroids, in which, similar to *Corymorpha nutans*, the planulae, by changes and budding, evolve into free swimming, small medusae from which colonial hydroids finally develop. True medusae, like *Pelagia noctiluca*, also belong to this phylum; they swim using their large bell and light up the sea on dark nights.

Sarcophyton sp.

The structure of cnidarians is more developed than that of sponges. There is a regional specialization within the tissue layers as well as a gastrovascular cavity; the symmetry of the external openings, in particular of the tentacles surrounding the mouth, reflects the internal axial symmetry.

CORAL REEF STRUCTURES

Coral reefs are peculiar habitats typical of warm waters. Although reef-building corals can tolerate temperatures of 18 °C, few reefs occur poleward of the 20 °C winter isotherm. Coral reefs are particularly well developed in an area extending approximately thirty degrees on both sides of the equator (Caribbean, Red Sea, northern Indian Ocean, Malaysia, Indonesia, New Guinea and northeastern Australia, and central Pacific) where the water temperature ranges from 25 to 29 °C. The west coasts of South and Central America, as well as the west coast of Africa are an exception; although they belong to the tropical zone, such coasts have nearly no coral reef owing to strong upwelling of cold water.

109

The development of coral reefs also requires that the sea salinity be within the normal range (32-35 grams per liter). The dilution brought about by the freshwater of the Orinoco and Amazon rivers curtails coral reef formation along the Atlantic coasts of South America.

The term coral reef encompasses the whole structure, including lagoons and off-reef calcareous deposits that play an important role in the life of the reef. Morphologically, the main types of coral reefs are the barrier reef, the fringing reef, and the atoll. The barrier reef develops parallel to the coast, from which it is separated by a shallow lagoon. The largest barrier reef is the Great Barrier Reef,

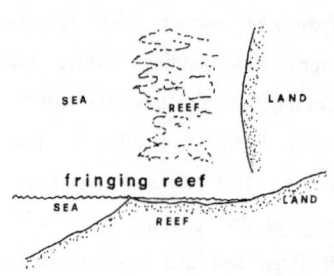

which is actually composed of many reefs; it extends from New Guinea to northern Queensland in northeastern Australia, over 2000 kilometers with widths ranging from 16 to over 350 kilometers. The second largest barrier reef extends over 640 kilometers along the northern coasts of New Caledonia; the third largest, the Great Sea Reef of Fiji, is nearly the same size.

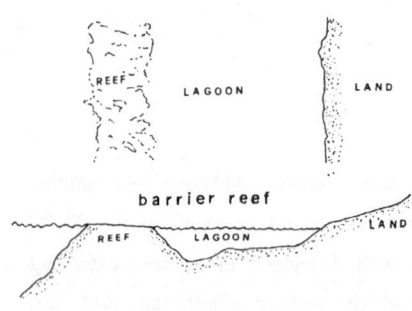

Fringing reefs develop close inshore on rocky coastlines; they grow best where no sediment is transported from the land, i.e. where there are no rivers, and the land is arid. The longest fringing reef, favored by these conditions, extends along the Red Sea coasts over 4000 kilometers, ideally straightening the shores.

CNIDARIANS

The atoll is an annular coral reef which encloses a shallow lagoon with islets. It is typical of the Indo-Pacific area, where it can develop from the deep sea as an isolated structure. The largest atolls, with diameters exceeding 29 kilometers, are the Suvadiva Atoll in the Maldives and the Kwajalein Atoll in the Marshall Islands. Atolls can also develop near the coast from the continental shelf.

A minor type of coral reefs is the patch reef, which is a very common structure which develops from shallow waters in lagoons and submarine shelves. At low tide, its top flat of dead corals breaks the surface of the water causing problems to outboard motors. However, one can tell when one is approaching a patch reef by the change in the color of the waters from bluish to yellowish; only in turbid waters such as those in Kanehoe Bay in Hawaii during the rainy months, did I have problems on the patch reef with my Boston Whaler.

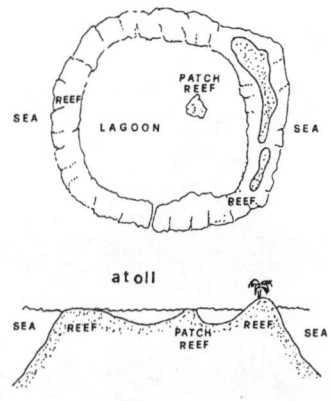

Lagoons of both barrier reefs and atolls have also isolated mounds of living corals called coral knolls, faros (or atollons, which are small-sized atolls), and microatolls (which are smaller formations of the Great Barrier and the Caledonian coral reefs). Finally, table reefs are isolated structures that develop in the Caroline and Marshall Islands from deep or intertidal waters where there is no lagoon.

The solid structure of coral reefs is mainly due to both scleractinian corals and encrusting coralline algae. *Acropora* is the prevalent genus of coral in warm waters; in Indo-Pacific areas it comprises many species which typically grow together, whereas in Atlantic areas there are fewer species. Also important as reef builders are the genera *Pavona*, *Goniopora*, and *Pocillopora*, which are not found in

Atlantic areas. In Hawaii, *Porites compressa*, *Pocillopora damicormis*, *Montipora verrucosa*, and *Fungia compressa* are the dominant genera, while *Acropora* is absent, or only present in western islands as an atypical, non-reproductive species.

The third most important coral reef builders are the hydrocorals, mainly of the

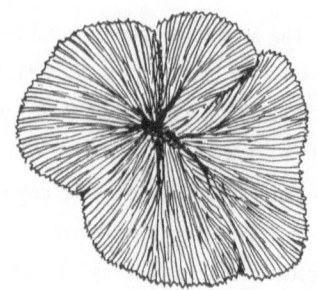

Fungia compressa

Tubipora musica

genus *Millepora*, followed by the calcified octocorals. The firmest of the latter is the blue coral, *Heliopora coerulea*, an octocoral which forms parallel solid tubes in warm waters. More commonly encountered is the octocoral *Tubipora musica* (organ-pipe coral), which has a less solid tubular and spicular structure.

Coral reefs are also populated by alcyonarians, particularly in northern Indo-Pacific areas; these soft octocorals are sometimes dispersed in a matrix of sponges and therefore can not contribute to the solidity of the reef.

Atlantic coral reefs are richer in gorgonians than Indo-Pacific reefs, whereas in Hawaiian shallow waters the gorgonians are rare and the alcyonaceans absent.

The fauna of coral reefs is as diverse as one can imagine, and not only

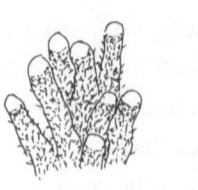 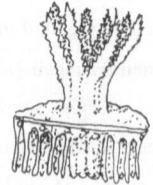

Heliopora coerulea
(right: polyp and solenial tube)

in regard to scleractinians and other corals. Corals and seaweeds compete for space and food, with a general prevalence of corals; the seaweeds are grazed and the corals are preyed upon and dead portions of the reef are destroyed by both waves and boring animals. Unbalancing this equilibrium by causing either a high a death rate or overproduction of algae, would have catastrophic consequences for the reef.

HISTORY OF CORAL REEFS

In 1859, before presenting the theory that man might have evolved from lower forms of life, Darwin explored the coral reefs. Perhaps this served as the most fruitful of all scientific marine cruises, as it revealed the fundamental history of coral reefs which is contemplated in the Subsidence Theory. According to Darwin, fringing reefs were first formed along the coasts of volcanic islands in the Indo-Pacific. Then, an earthquake and the subsequent rising of the earth elsewhere brought about the subsidence of these high volcanic islands, and fringing reefs were submerged while they continued to grow forming barrier reefs. Further subsidence led from barrier reefs to atolls.

Darwin's theory was strongly disputed by the scientific world until it was definitively proven in 1951 by deep drilling at Eniwetok Atoll in the Marshall Islands. This atoll proved in fact to consist of over 1200 meters of coral limestone dating back to the Eocene era which lay at the top of a basalt volcano rising over 3300 meters from the ocean floor. Drilling in southern New Caledonia and on Heron Island provided further evidence for the Subsidence Theory.

I do not mean that all types of coral reef must have formed the same way, however. Less ancient surface reefs may have started to form in tropical waters during the low sea level periods in the Pleistocene glaciation. With the dissolution of ice and the subsequent rise in sea level, corals continued to grow and form barriers. This appeared in a more restricted formulation as the Daly's Glacial Control Theory; it does not oppose the Subsidence Theory as the areas of application of the two theories are quite different.

Strictly speaking, Daly's theory may only apply to Indo-Pacific reefs as the Atlantic Ocean became colder than the Indo-Pacific oceans during the glacial periods, and coral growth was substantially inhibited. Corals could only start to grow vigorously again in the Atlantic Ocean at the end of the last glaciation. They grow now more vigorously and deeper than the corals of Indo-Pacific reefs. This implies that Indo-Pacific coral reefs are older than Atlantic coral reefs. Whereas the latter are "only" 10,000-15,000 years old, some Indo-Pacific reefs, particularly some atolls, are the products of several millions of years of coral growth. A prospect that would have disgusted archbishop Ussher.

The most recent, and still scarcely developed, coral reefs may have started to form on platforms of various origins, with no involvement of either glaciation or volcanism. This is the Antecedent-Platform Theory of Hoffmeister and Ladd which has found support in the successful transplantations of corals which formed new patch reefs such as those of Kanehoe Bay in Hawaii.

THE FEEDING OF REEF-FORMING CORALS AND THEIR PROTECTION AGAINST SUNBURN: THE STORY OF THE ZOOXANTHELLAE

An important aspect of reef building corals is their association with photosynthetic dinoflagellates (zooxanthellae, mainly *Symbiodinium microadriaticum*). It is a phototropic symbiosis of the type already discussed in the association of cyanobacteria with sponges. There is a direct vital translocation of nutrients from the zooxanthellae to the coral host, which is responsible for the enormous success of corals in coral reefs.

Zooxanthellae-coral systems need light, but not a high energy light, otherwise they would burn the same way fair-skinned northern Europeans do on a tropical beach. In chapter 2 I have stressed the importance of solar light for the photosynthetic process of the zooxanthellae in coral reefs: in fact, it is the medium-energy light of the visible spectrum which is utilized in the photosynthetic process. Solar light of shorter wavelength, the ultraviolet light that is scarcely filtered by the ozone-poor atmosphere of the tropics and which penetrates the clear tropical waters, is strong enough to damage both the zooxanthellae and the corals.

The problem is that of filtering strong solar radiation. In most fair-haired people this is badly accomplished through the pheomelanin in their skin. The skin of black-haired people contains eumelanin, which turns brown in the sun and is a more efficient filter than pheomelanin. Similarly, to protect themselves and the corals against sunburn, the zooxanthellae developed enzymatic systems that elicit the production of pigments which function as solar-radiation filters. Thus, in *Acropora formosa* (staghorn coral) of the Great Barrier Reef special non-protein amino acids, like palythine, act as protecting pigments. This and other structurally related pigments of the coral have a wide distribution and apparently do not always function as solar filters. In fact, palythine has also been isolated from the tropical hexacoral *Palythoa tuberculosa* and related compounds have been isolated from both terrestrial fungi and the Mediterranean bivalve mollusc *Mytilus galloprovincialis*. In the latter case the role of palythine and similar amino acids can hardly be that of solar filters: the mollusc lives in temperate areas and is protected by a hard shell.

In addition to pigments, *A. formosa* and other reef-forming scleractinians produce a large amount of mucous polymer for protection against rapid desiccation in the sun at low tide. This mucous substance shows up as a white slime at the surface of the waters at rising tide.

Tropical scleractinians of the family Dendrophylliidae, mainly of the genus *Tubastraea* (daisy coral), which can not be considered as reef builders, contain different pigments, the aplysinopsins. These are alkaloids derived from the protein amino acid tryptophan. Though tropical dendrophylliids are not as exposed to sunlight as corals of the genus *Acropora*, certain aplysinopsins show the characteristics of solar filters. In fact, while all aplysinopsins absorb sunlight, certain ones have an additional mechanism to capture solar radiation, i.e. by undergoing reversible phototransformations. In any event, the dendrophyllids differ sharply from *Acropora*: while in *Acropora* the pigments are produced by zooxanthellae hosts, in the dendrophylliids pigment production by zooxanthellae has yet to be demonstrated and the aplysinopsin-type pigments are only found in species that lack zooxanthellae.

Cases are known of corals that can survive bleaching, i.e. loss of zooxanthellae and pigments when held in dark caves, and can recover their algae when exposed to the sun again. In general, however, corals devised to live in association with zooxanthellae depend completely on these algae, and bleaching is generally followed by rapid death. The same also occurs in other cnidarians and sponges in tropical areas. The phenomenon of bleaching, which probably starts with photochemical and thermal injuries to the zooxanthellae, is of high concern in west tropical Atlantic coral reefs.

Whether the zooxanthellae also play a role in the fixation of calcium is uncertain. What we know is that these algae provide the corals with other important metabolites besides pigments. One of such metabolites is dinosterol, as discussed in Chapter 2. Another case in which the zooxanthellae are probably involved is that of *Echinopora lamellosa*, a scleractinian coral of the family Faviidae found around Korolevu, in the Fiji Islands. Terpenes of this coral, e.g. (20ß)-echinolactone-B, are typical of plants and might therefore originate from the zooxanthellae hosts rather than from coral cells.

THE STINGING

The hydroids live in all seas and in freshwater, but are particularly abundant in the tropics where they contribute to the building of coral reefs, mainly in the genus *Millepora*.

In spite of their appearance as fragile flowers, most hydroids have powerful means for defense and offense in the form of cnidocysts with puissant stinging properties which are used to paralyze their prey. A beautiful such solitary species equipped with a rudimentary chitinous outer layer is *Corymorpha nutans*. All cnidarians have cnidocysts, which are not exactly cells, but derive from cells. Luckily, not all cnidarians can effectively sting man, however.

Stinging cells occur singly or in groups on the tentacles, around the mouth, or within the body. Two types of stinging cells are known, nematocysts (stinging cells which consist of a sort of horny hypodermic microsyringe) and spirocysts (which are non-penetrating stinging cells typical of sea anemones and which can only be used once to discharge the venom into either the prey or the unfortunate swimmer).

Corymorpha nutans

Swimming in waters populated by jellyfish and true medusae (Cubomedusae) which have long filaments covered with nematocysts may be unpleasant indeed. True medusae of the genera *Chironex* and *Physalia* in certain Australian, Asiatic, and

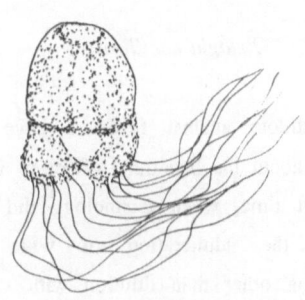

Chironex fleckeri

Caribbean areas, and *Chiropsalmus* along the Mississippi Gulf Coasts, constitute a public health problem. The "Irukandji syndrome", which results from stinging by such medusae, may disable man for several days, and many deaths caused by *Chironex fleckeri* have been recorded in northeast Australian waters. The venom, which has hemolytic and musculotoxic activities, consists of thermolabile proteins. Older drugs, either natural or synthetic, used to alleviate such problems, contain antihistaminic compounds, as in the treatment of allergies. It is good that the

117

Commonwealth Serum Laboratories Australia now produce antivenoms which are effective against *Chironex* stings.

The stinging cell capsule is made of collagenous proteins bound together by sulfur bridges. Although the problem has not yet been fully clarified, the venom of the nematocysts induces chemical reactions which differ from those of the spirocysts; this indicates a subtle difference in the characters of the sulfur bridges in the two types of stinging cells.

THE LUMINOUS

The emission of a luminescent blue slime by the Mediterranean jellyfish *Pelagia noctiluca*, the "pulmo marinus" already noted by Pliny the Elder and later studied by Spallanzani, is the only well-documented case of extracellular bioluminescence besides that of the crustacean *Cypridina hilgendorfii*. Until the mid-sixteenth century it was thought that such a blue secretion mixed with wine was

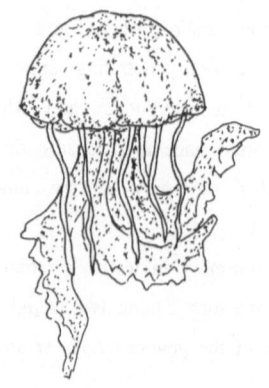

Pelagia noctiluca

an antidote against fever. I have no doubt about the beneficial effect of wine of that time, when technology did not allow the adulteration of wine by anything other than dilution with water (that was not polluted).

The blue light of *P. noctiluca* is peculiar since *Aequorea forskalea*, like most other luminous cnidarians, emits green light. The luciferin (aequorin)

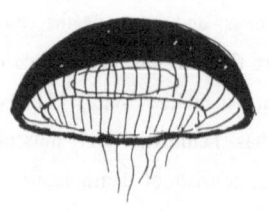

Aequorea forskalea

involved with *Aequorea forskalea* is a protein in which the active center is an aromatic alkaloid called coelenterazine; the process requires calcium ions, but no molecular oxygen, and includes another protein which undergoes excitation transfer from the luciferin and emits green light. This second protein is not present in *P. noctiluca* which therefore emits blue light directly from the luciferin.

There are other luminous cnidarians besides jellyfish. The luminescence of sea pens (related to the "penna marina", probably *Pennatula phosphorea*) is described in notes dating back to the mid-sixteenth century. The propagation of the luminescent waves throughout the colonies, in concentric rings from the center of stimulation, attracted great attention.

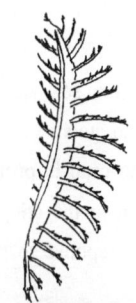

Pennatula phosphorea

Studies with luminous sea pens are generally more difficult than with jellyfish, since most sea pens settle on the sea bed, typically at a depth of 60-80 meters, which is below the limit for normal SCUBA diving. There are exceptions in certain areas, for instance *Veretillum cynomorium* in East Pyrenean waters. Although in such areas this sea pen is more abundant at a depth of 60 meters, or even deeper, it may be found at depths of less than 10 meters near Banyuls-Sur-Mer. This is due to the turbidity of east Pyrenean waters, as we remarked before, so that marine invertebrates in the search of light live closer to the surface than elsewhere.

Diving difficulties notwithstanding, the bioluminescence of the sea pen *Renilla reniformis* has been studied in detail: the luciferin-luciferase system, in contrast to that of jellyfish, requires molecular oxygen.

MACROFAUNA

NATURAL PRODUCTS OF CNIDARIANS IN PHYSIOLOGY AND PHARMACY

Sea anemones are mostly sessile organisms which settle on rocks and sediments or move slowly from place to place. Some of them smartly settle on shells occupied by active crabs, like *Calliactis parasitica* on the hermit crab. In this way the sea anemone benefits from a free shuttle and gets residual food from the crab. In return, the crab is protected from predators by the venom in the stinging cells of the sea anemone.

Calliactis parasitica
on the hermit crab

In response to strong stimuli, however, sea anemones may detach from their

Anemonia sulcata

supports and swim away by synchronized tentacular sweeps or by flexion of the column, a phenomenon called "swimming response".

The sea anemone *Anemonia sulcata*, which can be easily collected along northeastern Atlantic coasts, has acquired much importance in the area of physiology. Polypeptides of this animal strongly affect the sodium channels of cell membranes, a vital affair that we have already considered in regard to the dinoflagellate toxins (saxitoxin and gonyautoxins).

In what concerns cytotoxic metabolites, the telestacean octocoral *Telesto riisei* from Hawaii contains prostanoids, which are acetogenins structurally close to the prostaglandins and which have been called punaglandins from the Hawaiian word "puna" for "coral". Related prostanoids from

Telesto riisei (polyp)

the stoloniferous octocoral *Clavularia* sp. of Okinawan waters show similar, though weaker, cytotoxicity.

Certain fatty acids of the black coral *Leiopathes* sp., dredged at a depth of 290 meters around Saint Paul Island in the southern Indian ocean, are related to prostaglandin precursors. This is a most recent discovery in my laboratory as the natural product chemistry of black corals was totally unknown in spite of the importance of black corals. These colonial hexacorals are tree-like formations with a wide habitat; they can be seen on reefs when snorkeling but they also replace the sun-seeking scleractinians at the base of the coral reef and even inhabit deeper waters, (e.g. *Leiopathes* sp.) where they grow older and thus more solid.

Inhibition of cell division of fertilized eggs of sea urchins and starfish, often accompanied by cytotoxic activity, is worth considering here. A case in point is that of a gorgonian of the Gulf of Suruga in Japan, *Acalycigorgia inermis*, whose active products are terpenes which have been called acalycixeniolides for their structural resemblance to metabolites of alcyonarian soft corals of the genus *Xenia*.

Zoanthoxanthin, an unusual pigment of the Mediterranean zoanthid *Parazoanthus axinellae* and of related tropical genera, and structurally related compounds of this and other zoanthids, have elicited much interest for their ability to intercalate into DNA. No pharmacological application has been suggested, however.

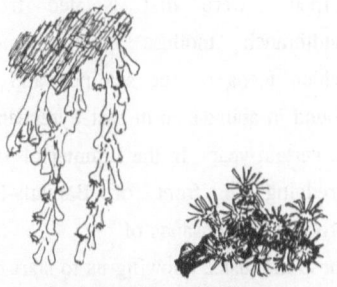

Parazoanthus axinellae

As to the vascular activities, I can mention various examples; polypeptides of *Anthopleura elegantissima* and *Anthopleura xanthogrammica*, sea anemones which can be easily collected along the Pacific coast of the USA, exert a cardiotonic effect by affecting the translocation of calcium ions. Atriastimulant action, though less intense than with dopamine, is triggered by products of two corals of the Great

Barrier Reef; the alcyonacean *Sinularia flexibilis* contains an active nitrogenous compound while the sea pen *Scytalium tentaculatum* has an active terpenoid structurally related to renillafoulin-A of the sea pen *Renilla reniformis* which is described below. In the same area one can find the alcyonarian soft corals *Sinularia heterospiculata* and *Nephthea* sp. which both contain dopamine.

Neuroactive substances have also been found in sea pens. The responsible compounds are diterpenes of a type first found in the gorgonian *Briareum asbestinum* of the Caribbean, and thus called briareins. They are structurally similar to the compounds of sea pens, for instance renillafoulin-A. These compounds affect the central nervous system, but their poor solubility in water and thus in body fluids is a drawback to their medical use.

The terpenes of the Mediterranean sea pen *Veretillum cynomorium* are related to these compounds, both in regard to molecular structure and bioactivity. These terpenes were first isolated from the nudibranch mollusc *Armina maculata*, which feeds on the sea pen and can be found in abundance in east Pyrenean waters in certain years. In the summer of 1982 our dredging in front of Banyuls-Sur-Mer brought in thousands of these mollusks, allowing us to start our work.

Veretillum cynomorium

Scleractinians of the family Dendrophylliidae, including some species of *Tubastraea* both from the tropics and from temperate areas, contain an alkaloid which strongly affects neurotransmission, acts as an antidepressant, and alleviates the noxious side effects of certain drugs currently in medical use. This alkaloid, methylaplysinopsin, which is one of the solar filters in the above-mentioned story of the zooxanthellae, was first isolated from two sponges of the order Dictyoceratida:

Aplysinopsis reticulata of the Great Barrier Reef and *Smenospongia* [= *Polyfibrospongia*] *echina* of the Florida Keys.

Lophotoxin, a terpenoid of sea whips of the genus *Lophogorgia* found off the Pacific coast of Mexico, is a special case. Though toxic, as the name implies, lophotoxin is a most useful neuropharmacological probe for the study of the nature and functions of nicotine receptors, to which, as a unique case, it binds irreversibly. If tobacco manufacturers and governments were aware of these properties, *Lophogorgia* depletion campaigns would certainly be launched.

Caffeine is another widely used drug, though a harmless or perhaps useful one in low doses. The Turks are known for their notoriously high daily dose of caffeine; and when I read that in a laboratory in Istanbul caffeine had been found in the gorgonian *Paramuricea chamaeleon*, it made me think of Turkish coffee spilling over the chemical bench.

Pseudopterosin-A, a substance of mixed terpene-sugar biogenesis, isolated from the sea whip *Pseudopterogorgia elisabethae* (sp. nov.) of the Bahamas as well as from other gorgonians of the same genus, has all the characteristics of a modern antiinflammatory drug. Although the mechanism of action is not yet clear, pseudopterosin-A does not mimic morphine nor antagonize histamine nor inhibit prostaglandin synthase; perhaps it functions as a structural mimic of phosphatidylinositol, thus affecting the production of icosanoids, inositol triphosphate, and diacylglycerol.

Cladiellin, a terpenoid of the soft coral *Cladiella* sp. of the Great Barrier Reef, has strong antiinflammatory properties too, but, unfortunately, it is toxic. The litophynins are metabolites structurally similar to cladiellin; they have been isolated from the soft coral *Litophyton* sp. of Sukumo Bay in Japan and show insect growth inhibitory action, specifically against the silkworm (*Bombyx mori*).

Similar properties are observed in another, structurally related terpene called sarcodictyin-A, which was isolated from the stoloniferous octocoral *Sarcodictyon roseum* [= *Rolandia rosea*]. This smooth Mediterranean coral settles on the skeleton of gorgonians, often together with the alcyonacean *Alcyonium* [= *Parerythropodium*] *coralloides* which is as coarse as a shark's cuticle; actually, the smooth/coarse traits serve to distinguish the two corals underwater. *S. roseum* is by far the less abundant of the two corals and I only found it in sufficient amounts for my studies in east Pyrenean waters.

Antimicrobial activities have not been investigated much. In one of the few studies, the hydroid *Garveja annulata* of Barkley Sound, British Columbia, proved to contain antimicrobial aromatic compounds which have been called garveatins.

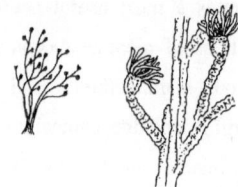

Garveja annulata

ECOLOGICAL ROLE OF NATURAL PRODUCTS OF CNIDARIANS: *Gerardia* AND *Laura* REVISITED

Researchers investigating the ecological role of marine metabolites are faced with the difficulty of reproducing a natural environment in bioactivity tests. As a consequence, the majority of the marine natural products that have been isolated have not been tested at all for their ecological relevance. Therefore, once the methods have been refined, it is likely that the ecological significance now ascribed to some marine natural products will be refuted.

Most of what we know about the ecological role of natural products of marine cnidarians dates back to studies undertaken in the seventies in the United States. Marine biology has received little support since then, so that studies of marine natural products in the United States have had to consider pharmacological properties to get financial support.

Other countries, such as Japan, with a strong tradition in marine research, are having their problems too and immediate economical problems dominate research;

edible crustaceans and mollusks are looked upon more favorably than cnidarians.

When one recalls such a successful enterprise as the dating of oil shales through the dinoflagellates (Chapter 2), one can appreciate that neglecting marine biology may be an error even in terms of immediate reward.

In any event, for my readers the cnidarians are certainly attractive organisms, for instance for their defense systems. Gorgonians, alcyonaceans, and sea pens have developed defense systems against fouling organisms. This applies to the branched gorgonian *Leptogorgia virgulata* and the unbranched gorgonian *Leptogorgia setacea*, which are commonly found in subtropical, Atlantic coastal areas of the United

Leptogorgia virgulata

States. These and other gorgonians secrete an amino acid, homarine, which has antifouling properties, specifically against the diatom *Navicula salinicola*. Both the gorgonian *L. virgulata* and the sea pen *Renilla reniformis*, collected in Sagami Bay, Japan, contain low molecular weight polymers which inhibit the settlement of barnacles. Another case is that of the sea pen *Renilla reniformis* found along the

coasts of North Carolina. The active compound, renillafoulin-A, inhibits the settlement of larvae of the barnacle *Balanus amphitrite*. Similar compounds are contained in the Mediterranean sea pen *Veretillum cynomorium* and in the nudibranch *Armina maculata* which feeds on it.

Balanus amphitrite

The gorgonian *Muricea fructicosa* found

along the coasts of California contains saponins, called muricins, which keep it free from fouling organisms. Another species of the same area, *Muricea californica*, for unknown reasons does not contain any muricin and is therefore heavily fouled.

The gorgonian *Pterogorgia citrina* has set up a special antifouling mechanism which resembles that already discussed in certain seaweeds. The gorgonian secretes a mucous substance on which fouling microorganisms settle. Once the settling phase is completed, the mucous substance is ablated with its inhabitants.

The whip-like sea pen *Stylatula* sp. of Isla Partida in the Gulf of California contains a terpene, called stylatulide, which is extremely toxic to the larvae of the copepod *Tisbe furcata*.

A soft coral that has developed a defense mechanism against crustaceans is *Cespitularia* aff. *subviridis* of the Seychelles. When molested, this alcyonacean emits a mucous substance that contains a terpene, palustrol, which is toxic to crustaceans.

Before concluding this passage on products used against crustaceans, let me mention a striking fact: high-molecular-weight polymeric substances of the sea pen *Renilla reniformis* and of gorgonians of the genus *Lophogorgia* induce the settlement of barnacles whereas, as seen above, low-molecular-weight polymeric substances of these organisms oppose that settlement.

Certain alcyonaceans and gorgonians have managed to protect themselves against predator fish. The most complex case is that of the soft coral *Litophyton viridis* of Laing Island in the Bismarck Sea, off the northern coast of Papua. This coral has developed toxins (a mixture of unidentified terpenes) active against local carnivorous fish but harmless toward *Abudefduf leucogaster*

Abudefduf annulatus,

a damselfish

which, like all other damselfish, is herbivorous. The latter is thus protected, along with its garden, against other fish. If you remember the problems the other damselfish have on coral reefs, you may conclude that *A. leucogaster* is a lucky species indeed. But the story of *L. viridis* does not end here. A smart carnivorous fish, *Chaetodon ocellinatus*, which is only sensitive to the toxins of *L. viridis* through its gills, feeds on the soft coral by biting on one colony and then on another one at such a distance that the toxin that spreads out from the first bite can not harm it.

The coffee-brown, highly branched gorgonian *Plexaura homomalla* (Esper) forma *kükenthali* Moser is extraordinary in that it contains large amounts (up to 2% of dry weight) of the methyl ester of human prostaglandin-A2; the ester induces severe vomiting in fish. In spite of this, the snail *Cyphoma gibbosum* feeds with impunity on the gorgonian and takes advantage of its distasteful compounds as a protection against predators.

Less complex, as far as we know, are the gorgonian *Pacifigorgia* cf. *adamsii* of Bahia Los Frales in California and the soft coral *Lobophytum denticulatum* (Tixie-Durivault, 1956) of Okinawan waters. Their toxins, the terpenes pacifigorgiol and denticulatolide, are strongly toxic respectively to the damselfish *Eupomacentrus leucostictus* and

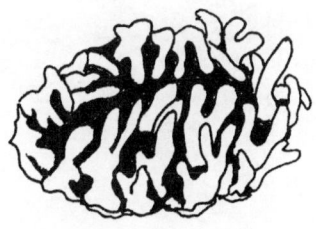

Lobophytum denticulatum

the brackish-water fish medaka (*Oryzias latipes*). Denticulatolide has a molecular structure of the type of lophotoxin, which is a toxin found in gorgonians of the genus *Lophogorgia*, as described below.

Tropical waters, particularly Atlantic coral reefs, are the typical habitat of gorgonians. But giant gorgonians of the suborder Scleraxonia, which look like enormous alcyonarians, populate the coldest waters, too. It was during a French

cruise aboard the Marion Dufresne in the
pre-Antarctic waters of the Crozet Islands
in 1982 that the giant gorgonian
Paragorgia arborea was collected by
dredging at a depth of 270 meters. A
terpene, arboxeniolide-1, of this gorgonian
is similar to terpenes of another
scleraxonian, *Corallium* sp., a precious
coral found in Hawaii; and the two
gorgonians also contain identical terpenes.

Paragorgia arborea
(portion of)

Inhibition of cell division of fertilized starfish eggs may also be considered as

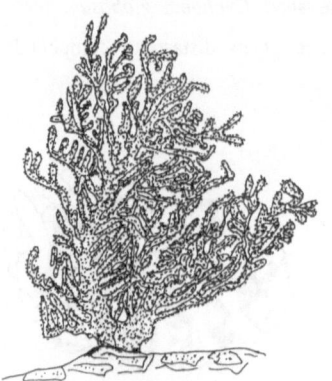

Euplexaura sp.
of Morito Beach

a defense strategy. Thus, the gorgonian
Euplexaura sp. secretes a terpene called
moritoside (the gorgonian was collected
at Morito Beach in Sagami Bay, Japan)
which inhibits the first cell division of
starfish *Asterina pectinifera* eggs; the
same occurs with saponins of another
gorgonian of the same area,
Anthoplexaura dimorpha. It is curious
that the active compounds are saponins,
which are the typical offensive products
of starfish and are contained in starfish
eggs, too.

A special case of chemical protection is encountered in the Mediterranean
zoanthid *Gerardia savaglia* (Bertolini, 1819). While most zoanthids are flat, *G.
savaglia* is arborescent with a vigorous skeleton of scleroproteins. The zoanthid and
a crustacean parasite host are at the center of a long story that begins with Henry de
Lacaze-Duthiers, a zoology professor at the Sorbonne in the last century. Working

together with professional red coral fishermen (*G. savaglia* has the same habitat as the red coral) the famous marine zoologist discovered the parasite on the apical portions of the zoanthid and named it *Laura gerardiae* after one of his relatives. It turns out that the apical portions of *G. savaglia*, where *Laura* is harbored, are more fragile than red coral and are therefore always the first to be broken when the red coral is collected with cables and nets bound to the heavy cross-shaped tool which was in use at that time. Areas of intact *G. savaglia* were consequently reputed to be virgin areas for red coral and the fishermen were unwilling to reveal their fishing spots to the zoologist who then abandoned the study of *Laura*.

Since then SCUBA divers have depleted this zoanthid, by collecting it as an ornament from Italian and French coasts. But it was in a tepid morning in early November that we accidentally discovered in southern Italy the last untouched refugee of *Gerardia*. Studies of this collection led surprisingly to the isolation in large amounts of ecdysterone, a hormone responsible for the molting of crustaceans. A novel steroid of the same

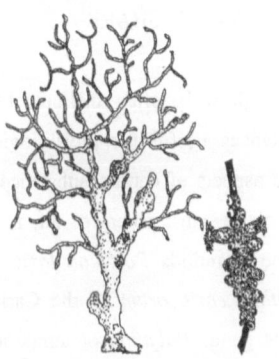

Gerardia savaglia

class was also found in this zoanthid and it was named gerardiasterone. *G. savaglia* is unique in the animal kingdom in that it contains zooecdysteroids in large amounts. In this respect, *G. savaglia* resembles certain terrestrial plants which also contain ecdysteroids, although of different structure and called phytoecdysteroids, in large quantities.

The idea that ecdysteroids can protect terrestrial plants against insects or *G. savaglia* against crustaceans by eliciting the molting process at a wrong time, is generally accepted, but it might be nothing more than a misleading anthropomorphic reasoning. Actually, the ecdysteroids are not merely molting agents but also powerfully bioactive products which act at the gene level. While their accumulation in our zoanthid may result from the lack of molt-inhibiting hormones (which are present in crustaceans and insects), understanding their role in our zoanthid may not be easy.

Sadly, there was no trace of *Laura gerardiae* in our various collections of *G. savaglia* and we were left with only the fine drawings of the crustacean reported by de Lacaze-Duthiers.

VENOMOUS CNIDARIANS: THE LEGEND OF THE CREATURE PARTLY MAN AND PARTLY SHARK

Although it is often difficult to make a distinction between venoms, toxins, and substances useful for physiological studies, I want to place emphasis here on the toxic aspects of certain substances of cnidarians. The most potent of such toxins, and one of the most potent of all, is palytoxin, a high molecular-weight acetogenin found in the zoanthids *Palythoa toxica* of Hawaii, *Palythoa tuberculosa* of Okinawa, and *Palythoa caribeorum* of the Caribbean. In contrast, palythoxin is absent from corals of the genus *Palythoa* of temperate waters, such as the coasts of Senegal.

The story of the Hawaiian species begins with the Tahitians who colonized the island of Maui after sailing the many thousand miles that separate Tahiti from Hawaii. At that time there was a suspiciously acting man on Maui and when the islanders discovered that he was half man and half shark, he was thrown into a pond close to the ocean where

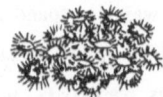

Palythoa tuberculosa

a reddish-colored moss grew. The moss (limu) became highly poisonous and the place was considered a taboo by the islanders. And as with any respectable taboo, anyone violating it was supposed to die.

CNIDARIANS

The poisonous pond was discovered in the sixties in Muolea, a district of the town of Hana on the island of Maui, and the reddish moss proved to be a zoanthid coral, not a moss. *Palythoa toxica*, as the zoanthid is called, is an euryhaline species which lived under less than ten inches of water in a pond in an area subjected to heavy rainfall. It grew only there, as far as one knows, and when I asked to visit the place, I was told that it had been destroyed by a recent hurricane; not foreseen in the legend, this was written in the indecipherable book of life.

In Okinawa it was noticed that the fish *Alutera scripta* was sporadically lethal to the pig, when given as food. Remains of *Palythoa tuberculosa* were detected in the gut of this fish so that a study of the toxin of this species of *Palythoa* began parallel to studies of *Palythoa toxica* in Hawaii. Remarkably, the Okinawan and the Caribbean zoanthid have a seasonal toxicity cycle whereas the species of Hawaii proved equally toxic throughout the year. Palytoxin and similar toxins of bacterial origin are found in other organisms, too; e.g. in the crab *Demania reynaudii* found in the Philippines.

The structure of palytoxin, which is the same from all sources, is so complex that long studies were required both in Japan and in Honolulu, and it was only through the total synthesis of fragments of the molecule at Harvard that all molecular details could be defined.

Palytoxin is a most potent systemic vasoconstrictor; at extremely low doses it shows reversible cardiotonic effects that begin with coronary constriction and reduction in coronary flow and extend to the vasculature with an increase in blood pressure. The action is peripheral, does not involve the inside of the heart, and at slightly higher doses systolic cardiac arrest occurs in a few minutes.

Subergorgic acid is a cardiotoxic terpene isolated from the gorgonian *Subergorgia suberosa* of Guam.

Remarkable is the effect of lophotoxin, a terpene found in sea whips of the genus *Lophogorgia* (*chilensis, cuspidata, panamensis*, and *rigida*) of Pacific Mexico. Lophotoxin acts by blocking neuromuscular transmission and prevents nerve

131

stimulated spasms; the symptoms preceding death, ataxia, paralysis, and respiratory depression resemble those of curare, a toxin found in terrestrial plants and used by the indigenous people of Amazonia to catch and to kill.

Structurally close to lophotoxin are the coralloidins, terpenes of the alcyonacean *Alcyonium* [= *Parerythropodium*] *coralloides*. This cardinal red soft coral is found in abundance in east Pyrenean waters, often together with the stolonifer *Sarcodictyon roseum*, on the skeleton of gorgonians of the genera *Eunicella*, *Paramuricea*, and

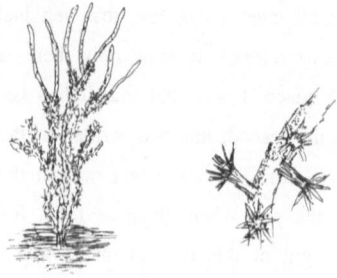

Parerythropodium coralloides

Lophogorgia; the gorgonians die and their skeleton is fully invaded by *A. coralloides*. The coralloidins are possibly responsible for this action.

8 Mollusks

The phylum Mollusca is a large one comprising about 75,000 species, mostly in one class, the Gastropoda. Mollusks are not only many in number but they are ubiquitous, too. They vary in size, from extremely small species to either the giant clam *Tridacna gigas* (which resembles *Tridacna maxima* but it may weigh up to a quarter ton) or giant squids. Although many groups of mollusks are exclusively marine, with habitats ranging from the intertidal to the abyssal zone, there are freshwater and terrestrial groups as well, and some species are endoparasitic.

Their fossil record is very rich and shells of mollusks are common on sandy beaches. Their forms, which follow precise geometrical laws, attract many collectors and are dealt with in innumerable popular books on shells. Ancestral mollusks, however, were quite different; they lacked a shell and were vermiform. Their classes, Caudofoveata and Solenogastres, though not as rich as the Gastropoda, still exist.

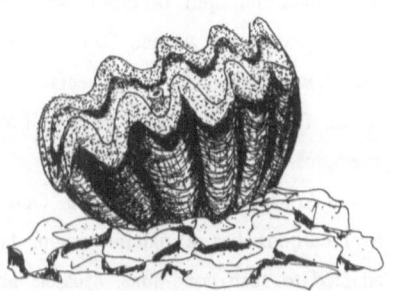

Tridacna maxima

We have already encountered mollusks that are capable of accumulating dietary toxins. This phenomenon is sporadic in the bivalve mollusks, as already discussed in Chapter 2: edible mollusks accumulate toxins of dinoflagellates and cyanobacteria by either filter feeding or grazing with their toothed tong-like chitinous radula, thus suddenly becoming a dangerous food. Other mollusks considered in Chapters 3 and 8 graze on seaweeds, like the sea hares, or prey on other invertebrates, like the sea slugs, and accumulate the bioactive products of their prey.

We have also encountered mollusks whose larvae were kept from settling by algal products.

MACROFAUNA

Octopus vulgaris, a common mollusc of the class Cephalopoda, is peculiar in that it contains a red iron-binding pigment, called adenochrome, whose structure has not yet been fully elucidated.

MUSSELS, OIL POLLUTION, AND DENTISTS

To run ships, jets, and cars, and to produce electricity, much oil had to cross the seas, and, unfortunately, some of it never reached its destination. It is not astonishing, therefore, that the sea bed is polluted in many parts of the world. An unfortunate oil spill occurred along the coast of Brittany, causing the ecological catastrophe of the Amoco Cadiz. Another major disaster occurred in seagrasses and mangroves, along the coral reefs, east of the Caribbean entrance to the Panama Canal, following the spill of 8 million liters of crude oil. And these are but two examples.

At several spots in the Gulf of Mexico pollution of the sea bed by hydrocarbons has a natural origin, however, and certain marine organisms have adapted to such conditions. Mussels have chosen to live there, feeding on methane and other hydrocarbons that spring from the sea bottom. This special case involves certain bacteria; you may remember from chapter 4 about the extreme adaptability of bacteria to all life conditions: well, mussels of the Gulf of Mexico have acquired bacterial symbionts, whose ability to feed on hydrocarbons has provided them with additional nutrients.

Certain mussels have been known since ancient times for a kind of precious silk (byssus), called "Meerseide" (seasilk) in Germany. The best byssus was used by the Romans to make cloths, and even earlier by the Egyptians to make linen for mummies. Byssus is a secretion consisting of polysaccharides combined with proteins and comes from the large Mediterranean mussel *Pinna nobilis* which lives erect on the sea bottom. Regrettably, this mollusc has been largely depleted from Italian coasts by dredging and by pirating divers.

MOLLUSKS

A few years ago a biotechnological company disclosed that they had found a way to stimulate mussels to produce relatively large amounts of a polypeptidic secretion which from its description appears to be byssus. Reportedly this is the best glue for dental protheses. I have heard about the total synthesis of such adhesives although they do not appear to be currently used.

OF SHIPWORMS, MARINE BACTERIA, AND CLIPPERS

Another example for the adaptation of mollusks to various life conditions involves the decay of wood in seawater; here, too, bacteria are involved. Ever since the Phoenician began to sail the seas, commercial clippers came into use, the first fixed marine installations were built, woody constructions were set up in tropical Pacific during the Second World War, and wooden Dutch dikes started to give way, man has been faced with big problems concerning the decay of wood in the sea. Wood-eating shipworms, which were later recognized as mollusks and not as worms, were responsible for such catastrophes. Today we are only left with a few relics of the romantic clippers, and their successors, dirty metallic tanks, have Iranian mines to fear, not shipworms. These mollusks harbor in a cellulose-digesting gland a bacterium which could have been devised by a science-fiction writer. The bacterium is capable of cellulolytic activity, which results in the degradation of the cellulose in wood, and of nitrogen fixation, which provides its host with elaborated nitrogen compounds. This case is unique in that the bacterium grows in the mollusc's gland in pure culture, ruling out the casual commensalism of most microbial symbionts of marine animals.

In addition to these special abilities, all bacteria benefit from the genetic organization of the prokaryotic cell which is best suited to meet their adaptational requirements. We know that the prokaryotes have extrachromosomal genetic elements called plasmids which code for catabolic activities and for drug resistance. Bacteria can eliminate or activate such functions in the plasmids without affecting other

135

functions, and this has nothing to do with genetic mutations or selection. It is a simple mechanism that bacteria can easily activate to endure extremely different life conditions.

WHEN EDIBLE SHELLFISH BECOME TOXIC

Certain mollusks, particularly bivalves, many of which are culinary delicacies, are excellent filter feeders. They can be dangerous, however, if there are toxic tiny organisms or toxic substances around, which may be retained in their organs or tissues causing them to suddenly become toxic. Take, for instance, paralytic shellfish poisoning which is due to saxitoxin and gonyautoxins of dinoflagellates, or ciguatera which is also due to a dinoflagellate toxin, although ciguateric intoxications mainly occur with fish.

The list does not end here. Diarrhetic shellfish poisoning is an unpleasant sickness caused by the toxins of dinoflagellates of the genus *Dinophysis*; the responsible non-blooming species in Japan, Spain, France, and the Netherlands were *D. fortii* and *D. acuminata*, while in Chile it was *D. acuta*. In the summer 1989 such phenomena appeared for the first time also in the Adriatic sea.

5 μm

Dinophysis sp.

The responsible toxins which accumulate in scallops and mussels are acetogenins called dinophysistoxins (structurally similar to okadaic acid), pectenotoxins, and yessotoxin. The latter is structurally related to brevetoxin-B of the dinoflagellate *Gymnodinium breve*.

In 1965 an outbreak of shellfish poisoning occurred in Japan due to the Japanese ivory shell (*Babylonia japonica*). The toxins involved are alkaloids, called surugatoxins, and are not biosynthesized *de novo* by the mollusc. A gram-positive bacterium of the Coryneform group, responsible for production of these toxins, has been isolated from the digestive gland of *B. japonica* collected in Suruga Bay.

Mussels cultured in Cardigan Bay, Prince Edward Island, Canada, were the cause of human poisoning in November 1987 with symptoms of amnesia. The poison was identified as the alkaloid domoic acid that we already know as an anthelmintic product of the red seaweed *Chondria armata*. In this case, however, domoic acid must come from a different source: these mussels, being filter feeding animals which are cultured by the hanging method, may well obtain domoic acid from plankton.

TYRIAN PURPLE AND OYSTER PEARLS: EXCLUSIVE PAST AND PRESENT COMMODITIES FROM THE SEA

Tyrian Purple, extracted from marine mollusks, was the pigment used by an exclusive Phoenician dying industry in the city of Tyre as early as 1600 years BC. Vast mounds of the remains of the mollusks accumulated around Tyre and Sidon, where the waters were reputed to give the best dye. North of Rhodes the dye was violet, while still more to the North it was almost black. Dyes of different colors were mixed to obtain different tonalities; and new towns were founded by the Phoenicians in search of new grounds for the mollusks.

The dye had already been known to the ancient Egyptians, as one can see from a red dyed shroud which at one time covered a mummy and which is now on display in the Kunstmuseum in Vienna. And, according to the Bible, the dye was used both in Babylonia (Jeremiah, **10**, 9) and for the hangings in the temples of the Israelites (Exodus, **25**, 4).

The reason for so much appreciation of Tyrian Purple was its resistance to strong sun, whereas vegetable red dyes and even cochineal red dye which were used by the Romans fade in the sun.

Tyrian Purple was derived from prosobranch mollusks of the genus *Murex* (*Purpura* to the Romans). The peoples of antiquity were already aware of the fact

137

that the mollusks do not contain the dye as such and that exposure to the sun is required for it to develop. That the dye is formed from colorless compounds contained in hypobranchial glands of the mantle was demonstrated by de Lacaze-Duthiers in the middle of the nineteenth century. Later one discovered that after an enzymatic step which does not go further than the green color stage the purple dye is formed in a photochemical reaction. And the German word "Purpur" seems to derive from the Indo-Germanic "bharbhur" which indicates something that changes or fades.

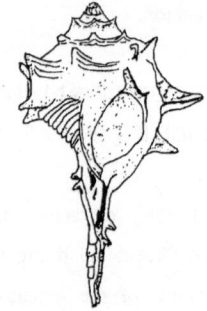

At the end of the seventeenth century, William Cole, a British master in the art of dyeing, being aware of the fact that the formation of red dye requires sunlight, modified the technology at the time by treating linen and silk with *Murex brandaris* extracts in the dark and then putting the fabrics under the sun where they turned purple.

Murex brandaris

The extremely small yield from mollusks made the price of Tyrian Purple higher than silver in Asia and only affordable to rich people and politicians (i.e. the same people). Thus, in ancient Rome Senators and dignitaries were called Purpurati and they had the exclusive right to wear purple. But there were so many of them during Rome's decadent period that a hill, Monte Testaccio, was formed by the remains of the mollusks. Purple was also the color of the robes of cardinals, until Pope Paul II had it changed to scarlet in the middle of the fifteenth century.

This exclusive ritual changed abruptly at the beginning of our century following one of the first ventures in marine natural products chemistry. At that time the German chemical industry was first class -this is still true today- combining science with economics. A case in point was the university professor Friedländer, who had extracted 22 grams of Tyrian Purple from 12,000 specimens of *Murex brandaris.* He discovered that the pigment was a brominated alkaloid (6,6'-dibromoindigotin) related to the protein amino acid tryptophan. He also devised an efficient method of total synthesis, adaptable to the non-brominated analog of the natural dye which made the dye cheap. And the myth of Tyrian Purple vanished.

Pearl oysters are an even older commodity, and one that has resisted the times. Already 4,500 years before the Christian era, the Hindus and the Chinese used oyster pearls as ornaments and the fisheries of the Arabic Sea are mentioned in the cuneiform inscriptions of Ninive. Pliny the Elder, following the Hindu legend that oyster pearls are concrete rays of the aurora, believed that oyster pearls were only beautiful under cloudless skies.

Pearl oysters are hard aragonite secretions formed as a defense against either a parasite or a sand grain in bivalve mollusks of various genera; true pearls are only found in the mantle of the pearl oysters, *Pinctada margaritifera* and *Pinctada mertensi*, however. Other tissues of the oysters may produce pearls, too, but of less value; they are called "perles baroques" in Paris. The best pearls have been fished in Sri Lanka, the Arabian Gulf, and the Red Sea.

ALGAE AND CYANOBACTERIA IN THE DEFENSIVE SYSTEM OF SEA HARES

Gastropods of the order Anaspidea are called sea hares because of the form of their rhinophores. They live in algal areas of very shallow waters, can only swim slowly, and have an internal shell or lack one completely. Their bodies are fleshy and vary in size; some species of *Aplysia* that live in tropical waters can grow to nearly half a meter in length and weigh up to two kilograms.

In spite of their fleshy constitution and slow movements, sea hares are not part of a marine predator's diet. The reason is that sea hares contain antifeedant or toxic compounds of seaweeds or cyanophyta, as we have already seen with the aplysiatoxins contained in *Stylocheilus longicauda*. As an exception, certain populations of the Indian Ocean do feed on fleshy opisthobranchs; the result is that intoxications by sea hares are common there.

Natural products may also be of endogenous origin in certain opisthobranchs; a case in point is aplysioviolin, a nitrogenous pigment which is distantly related to oxygen carriers and chlorophylls, and which is emitted by most sea hares of the genus *Aplysia* when molested.

Sea hares may also contain products with powerful pharmacological activity. For instance cyclic or linear peptides, called dolastatins, of *Dolabella auricularia* of the Island of Mauritius strongly inhibit proliferation of neoplastic cells. The most active of the peptides of *D. auricularia* is dolastatin-10 which has a linear structure. It strongly inhibits the proliferation of leukemic cells and of melanoma, and it is also active against viruses, such as *Herpes simplex* and equine rhinovirus.

Another significant case is that of aplysianin-A, a glycoprotein isolated from *Aplysia kurodai* of Japanese waters which induces lysis of murine tumor cells and has also antibacterial action.

Gastropods of the order Sacoglossa, which have the same habitat as sea hares, have the extraordinary ability to incorporate active chloroplasts from seaweeds into their diet and to keep them active. A good example is *Placobranchus ocellatus* which inhabits the southern coasts of Oahu Island in the Hawaii. Chloroplasts incorporated into this mollusc are the *loci* of synthesis of special acetogenins which, like tridachione, undergo photochemical reactions *in vivo* by absorbing the strong solar radiation of the tropics; thus, such acetogenins have the function of protective filters, as do the amino acids of reef-forming corals and the aplysinopsins of tropical dendrophylliids.

DIETARY NEMATOCYSTS AND OTHER CHEMICAL WEAPONS OF SEA SLUGS

Sea slugs are beautiful opisthobranchs of the order Nudibranchia. Their defensive system is mainly chemical, similar to that of sea hares. In general sea slugs incorporate defensive compounds from other invertebrates, predominantly sponges, hydroids, ascidians, and bryozoa, upon which they prey. As a special means of chemical defense, sea slugs may incorporate nematocysts from hydroids and other cnidarians. They also use physical methods of protection; notably the spicules of sponges that are part of their diet.

Nematocysts are used especially by nudibranchs of the suborder Aeolidacea which feed on cnidarians; the nematocysts are translocated through the digestive gland up to the dorsal cells where they become encapsulated and are maintained as a part of the mollusc. These sea slugs, in order to keep predators away, either hold up the dorsal portion where the nematocysts are located, or emit a cloud of exploding nematocysts.

Preying upon invertebrates which, like the sponges, may entail dangerous spicules or which, like certain sea squirts, contain a chemical as noxious as sulfuric acid, has required a great deal of adaptation by sea slugs. To tolerate such harmful constituents of their prey, sea slugs emit a mucus in which the spicules become embedded and the strong acid is incorporated. Nevertheless, spicules of sponges often penetrate the walls of the digestive apparatus of sea slugs.

The shape of sea slugs has become adapted to the prey: the doridacean sea slugs which feed on encrusting sponges are flattened and have a wide radula. In contrast, sea slugs which feed on erected organisms, hydroids for example, are smaller and more elongated, and, like certain aeolidaceans, have only a reduced radula, whose functions are replaced by strong jaws.

While the dorid sea slug *Discodoris planata* of the North Atlantic expels sulfuric acid for defense, other sea slugs use less drastic agents, usually antifeedant compounds. Thus, analog to the *Dysidea fragilis* in Chapter 6, the dorid sea slugs *Hypselodoris godeffroyana* and *Chromodoris maridadilus* of Kanehoe Bay on the Island of Oahu incorporate the antifeedant terpenes nakafurans from this sponge. Another dorid sea slug, *Phyllidia varicosa*, of Pupukea on the northern shore of Oahu, retains the antifeedant terpene 9-isocyanopupukeanane from the sponge *Hymeniacidon* sp. upon which it preys.

Dorid sea slugs of the genus *Tambje* of the Gulf of California use antifeedant, proline-derived alkaloids, called tambjamines, which they obtain from their prey, the bryozoan *Sessibugula translucens*.

Antifeedant compounds, though serving to keep most potential predators of sea slugs away, are harmless toward another sea slug predator. Thus, the large nudibranch *Roboastra tigris* preys upon *Tambje*, incorporates the tambjamines of the latter and becomes protected itself.

So far I have emphasized the dietary origin of antifeedant and toxic compounds contained in sea slugs. Frequently, however, the origin of protective compounds of sea slugs is not known; this is the case, for example, of antifeedant degraded steroids of the sea slug *Aldisa sanguinea cooperi* which are structurally similar to a repellent steroid used by the freshwater diving beetle *Dytiscus marginalis*.

In certain cases antifeedant compounds are synthesized *de novo* by sea slugs for their own defense and such compounds are located in the skin of the mollusc; an example is the Mediterranean dorid *Dendrodoris limbata* which contains the terpene polygodial, which is also an antifeedant compound of terrestrial plants. Another case is *Archidoris montereyensis* of British Columbia which synthesizes an antifeedant terpene belonging to the same class of 'caulerpatriene' found in green seaweeds of the genus *Caulerpa*. In a complex process, however, this dorid also uses dietary products obtained from the sponge *Halichondria panicea*.

Sea slugs can also make use of drastic protective organic compounds, such as verrucosin-A, an ichtyotoxic terpene of the Mediterranean dorid nudibranch *Doris verrucosa.* Other protective compounds of nudibranchs have already been mentioned in connection with sponges in Chapter 6. Latrunculin-A, for instance, is an acetogenin which disrupts the microfilamentous cellular organization; it was first identified in the Red Sea sponge *Latrunculia magnifica* but is also present

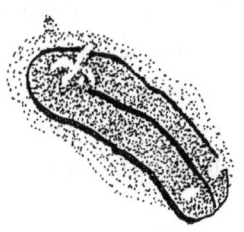

Chromodoris elisabethina

in the dorid mollusc *Chromodoris elisabethina* of the tropical West Pacific (both Guam and Eniwetok). Its origin, however, is still a mystery, several thousand miles away from the Gulf of Eilat where *L. magnifica* lives.

Sea slugs also contain products of potential pharmaceutical interest. A case in point is the acetogenin sphinxolide, isolated from an undetermined nudibranch found in Hawaii, and strongly active against tumoral KB cell lines. The name sphinxolide reflects the difficulty we had in our laboratory in defining the precise origin and, for some time, also the molecular structure of the compound.

Egg masses of nudibranchs have also proven productive. The egg masses of the dorid sea slug *Hexabranchus sanguineus* of Hawaii, for example, contain ulapualide-A, a compound whose name is derived from the Hawaiian words "ula" (red) and "pua" (flower) and which inhibits leukemic cells and pathogenic fungi. Another compound, kabiramide-C, obtained from egg masses of an unidentified nudibranch of Kabira Bay in Okinawa, has antifungal activity. Both ulapualide and kabiramide-C are macrolides which also embed nitrogenous functions probably deriving from amino acids.

DEFENSE AND DRUGS WITH PROSOBRANCHS

Most prosobranch mollusks are protected by a hard shell which is usually the only defense Nature provided. Certain prosobranchs, unsatisfied with only physical

protection, have developed chemical
defense strategies as well. The limpet
Collisella limatula of the Gulf of California
is a case in point, containing a terpene,
called limatulone, which repels crabs and
fish.

Certain prosobranchs are of potential
pharmaceutical interest: the common edible

Conus textile

clam with its antitumor activity, or abalone which contains antimicrobial and
antiviral proteins. However, the responsible compounds have only been adequately
described in a few cases, such as the hard-shelled snail *Kelletia kelletii* of the
Caribbean which contains special sugars, called kelletinins, with inhibitory action on
gram-positive terrestrial bacteria.

Conus textile, a hard-shelled mollusc of tropical seas, common in Okinawa, is
highly poisonous; there are records of human fatalities from its stings. The venom
has not yet been identified. We do know, however, that arachidonic acid is contained
in the venom duct, which ends in a venom-injecting apparatus at the tip of the
mollusc snout.

The toxicity of other species of *Conus*, such as *C. geographus*, is better
understood; the toxins, conotoxin and geographus toxin, are peptides.

THE AIR BREATHERS

The mantle cavity of certain mollusks is converted into a lung, which makes such
mollusks adapted for air respiration, without suppressing water respiration. Because
of this, these mollusks have been called pulmonates. Most species of this subclass
have freshwater or terrestrial habitats, like land snails; only a few species live in the
intertidal zone of the sea. Limpets of the genus *Siphonaria* are marine at all stages
and are therefore the best representatives of marine pulmonates.

MOLLUSKS

Interest in natural products from hard-shelled mollusks was first roused by pulmonates. Thus, *Siphonaria diemensis* of Phillip Island in Australia was shown to contain bioactive acetogenins, called diemenensins, which are structurally related to peroniatriol-1 of the onchidacean mollusc *Peronia peronii*. The diemensins inhibit the cell division of fertilized sea urchin eggs and also kill pathogenic bacteria, such as *Staphylococcus aureus*.

Siphonaria sp.

The onchidacean mollusks also inhabit the marine intertidal zone and are capable of respiration in water and in air; they have a controversial classification, either within the Opisthobranchia, or within the Pulmonata. These mollusks have no shell and are therefore particularly exposed to predators; they, in analogy to sea hares and sea slugs, had to set up a chemical defense mechanism. Thus, when molested, *Onchidella floridanum* and *O. binneyi* of the intertidal zone of the Bahia de los Angeles, secrete a liquid which acts as a deterrent against predatory fishes and crabs. The nature of the active principle is still unknown.

There are also compounds of potential pharmaceutical interest in onchidacean mollusks, e.g. *Peronia peronii* which is much appreciated as food in Micronesia and contains acetogenins, called peroniatriols, which are cytotoxic to leukemic cells.

9 Worms

WORM: A VAGUE TERM

In biological terms worm can be any of many animals belonging to different phyla, often with no more than bilateral symmetry in common. The turbellarians are a class consisting of mostly small and free-living marine species and belong to the phylum Plathyhelminthes which is the scientific name for flatworms. Though a revision is solicited by some zoologists, these are currently considered as the most primitive of the worms, although the identity of the ancestral turbellarian is uncertain.

Two other classes of worms descend from this mysterious ancestor: the flukes (Trematoda) and the tapeworms (Cestoda), both parasitic and mostly terrestrial, such as the pork tapeworm, *Taenia solium*. Freshwater and marine species of these two classes are also known, though they have been less investigated.

Proboscis worms, also called ribbon worms (the nemerteans), are carnivorous and mostly marine bottom dwellers found in shallow waters; they include extremely elongated species such as *Lineus longissimus*, which may attain a length of several meters when fully extended.

Roundworms (nematodes) are mostly small, free-living species that occur in large number in freshwater, in sea beds, and also in the soil; farm soil is made fertile by these worms. But there are parasitic roundworms as well, which attack all plants and animals.

Most people would hardly define the above, unsegmented, worms as pretty animals. Zoology textbooks show section drawings of dog kidneys invaded by the large nematode *Dioctophyma renale*, which is not exactly something you would want to look at during a meal. However, I am very respectful of the plans of Nature in designing living organisms, and I must admit that repulsion for worms is probably a distorting result of our education.

WORMS

In any event, there are cases of undoubtedly beautiful worms, too. Segmented worms (the annelids), which are mostly marine, comprise the fan worms *Spirographis spallanzani* and *Sabella pavonina* which are standard attractions in a marine aquarium. Here the trick is that the worm is hidden in a tube, and only the terminal part, which looks like a flower, is visible. And flowers are pretty things to anybody.

Spirographis spallanzani

Other annelids which move freely on the sea bottom have an enchanting name, *Aphrodita*, or by some authors *Aphrodite*, like the Greek goddess of sexual love and beauty. These annelids are beautiful indeed, their bodies sheltered in a mass of bright-colored hairs. That the name of these worms alludes to the sea is suggested by the worshipping of Aphrodite as a goddess of the sea and the seafarers. The Greek word *aphros* stands for "foam" and mythology tells us that Aphrodite rose from the foam that developed when Cronus threw the severed genitals of his father, Uranus, into the sea.

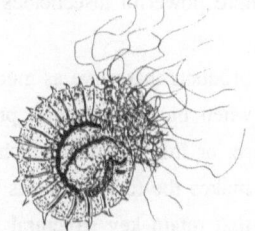

Amphitrite edwarsi

Certain sedentary annelids of the genus *Amphitrite* are also considered beautiful animals because their bodies are sheltered in a handsome tube and only hair-like tentacles are visible. Beautiful but aggressive: the sedentary *Amphitrite edwarsi* captures small prey with its long tentacles while the pirating errant worm *Lepidasthenia argus* quickly takes some of the prey away.

More evolved are the acorn worms, or hemichordates, a term which suggests that these worms have the characteristics of chordates. At the adult stage this is limited to the presence of gill slits; the supporting axis is different from that of the chordates. If we look at the embryonal stage, however, the hemichordates and chordates (and echinoderms, too) have many morphological points in common.

Still more evolved are the arrow worms, which belong to the phylum Chaetognatha of the chordates.

ANNELID INSECTICIDES, ACORN-WORM TUMOR INHIBITORS, AND PLASTICIZERS

What we know about the natural products of marine worms is largely confined to annelids and acorn worms. The annelid *Thelepus setosus*, of Kanehoe Bay on the Hawaiian Island of Oahu, contains an aromatic compound, thelepin, which bears structural similarity to antifungal products of the terrestrial lower fungus *Penicillium griseofulvum*.

Another annelid, *Lumbrinereis brevicirra* [= *L. heteropoda*], a common fish bait in Japan, inspired a rewarding venture. Starting from the observation that flies die in contact with this annelid, work in Japan led to the isolation of the active compound, a small sulfurated alkaloid which was called nereistoxin. This compound has served in Japan as a model to synthesize more powerful insecticides which are now on the market.

This is not an isolated example of natural products that serve as models for the synthesis of bioactive compounds. Particularly when bioactive natural products are not easy to obtain, from a rare animal for instance, or from animals which reproduce slowly, or where a complex molecular structure makes the total synthesis difficult, it can be easier to synthesize simpler compounds that retain key structural features of the natural product, and thus hopefully also its biological activity.

WORMS

One product, cephalostatin-1, is a steroid derivative which is markedly antileukemic. Despite the name, this compound is not contained in either cephalopods or fungi of the genus *Cephalosporium*; rather, it has been isolated from the small colonial hemichordate *Cephalodiscus gilchristi* (Pterobranchia) which lives in a tube in the Indian Ocean off southeast Africa. Interest in such a powerfully antileukemic agent as cephalostatin-1, and the need to have it in sufficient amount to clarify the molecular structure and to carry out screenings for biological activity, has prompted Americans to dive extensively at a depth of 20 meters to collect *C. gilchristi* in open seawaters controlled by the white shark.

Acorn worms belonging to the genera *Ptychodera* and *Glossobalanus* have been investigated for natural products. *P. flava laysanica* of Okinawa contains nucleosides of a type already found in ascidians and sea stars. A new, undetermined, *Ptychodera* sp. of the Island Maui in Hawaii, contains brominated aromatic compounds, and *Glossobalanus* sp. of Okinawa contains simple alkaloids.

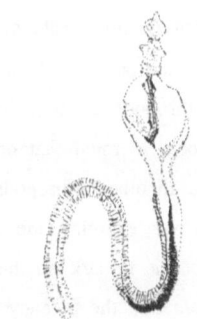

Ptychodera flava laysanica

The beautiful *Spirographis spallanzani*, however, can be full of pollutants. Specimens we collected in the harbor of Livorno in the Mediterranean proved to contain typical aromatic plasticizers in huge amounts. But this worm is not alone in this respect; lipidic foods, such as oil, butter, milch, yogurt, cheese, and so on, extract plasticizers from the plastic containers in which they are stored today. Plastic producers have prevailed against glass producers and we have been too weak in accepting plastics as food containers. I would be willing to pay more to have the food handled properly, but I am given little choice.

10 Crustaceans and chelicerates

Barnacles, shrimps, crabs, lobsters, wood lice, and crayfish are crustaceans, while horseshoe crabs, spiders, and scorpions are chelicerates. All of them, together with insects, centipedes and millipedes, belong to the arthropods which represent the vast majority of the animal world and have rich fossil records.

Edible crustaceans are of great economic importance; the culture of shrimps goes back long in time, as we have already seen in chapter 5 in connection with fungal infections. A new venture is the culture of spiny lobsters of the Caribbean; the trick in solving this problem was to collect the spiny lobsters from the sea with special nets at the post-larval stage.

The arthropods are considered an evolved stage of the annelids. In this regard, we have to bear in mind that only insects, centipedes, and millipedes have evolved on land while all other arthropods have marine origin.

Before we examine the role of chemical substances produced by marine arthropods, let me remark on the curious use of the transparent crustacean *Daphnia magna* in evaluating the potency of aphrodisiac substances; on this basis, aphrodisiac substances were in the recent past selected for use in animal breeding.

ADAPTATION AND DEFENSE OF BARNACLES, SHRIMPS, AND HORSESHOE CRABS

The crustaceans are mostly marine arthropods with some freshwater representatives. A large fraction of multicellular zooplankton in the oceans is made up of crustaceans, mainly the small copepods. The large lobsters, shrimps, and crabs, which are culinary delicacies, are crustaceans, too.

Barnacles can live on nearly any kind of substrate, including seagrasses, crabs, and turtles, but the most familiar ones are the intertidal species. Their calcareous shells with sharp edges create problems for people walking on the rocks in summer.

CRUSTACEANS AND CHELICERATES

Barnacles are zone-restricted, however, and more so in temperate than in tropical waters[8]; certain parts of the coasts of the English Channel, on both sides, are completely covered by barnacles whereas other areas nearby have none. Though settling factors have not been yet identified, the hatching factor, 'balanoacid', of *Balanus balanoides* (acorn barnacle) of Menai Strait in Wales, is a fatty acid structurally related to the precursors of the prostaglandins.

Other crustaceans, like shrimps, crabs, and lobsters, are covered by a hard, chitinous exoskeleton which helps locomotion and protects them from predators. The skeleton of many of these arthropods has a red pigmentation concentrated in small zones; the factor responsible for the concentration of the red pigment is a neuropeptide, i.e. a peptide secreted by neurons.

Because of its nature, the skeleton can not grow steadily with the animal; it has to be renewed at regular intervals during the animal's growth, after an inner cuticle has been set up. This process, called molting or ecdysis, is induced by steroidal hormones which, like ecdysterone, are called molting hormones or ecdysteroids. These compounds, which have unique structural features, are responsible for the molting of insects, too, and only appear during the molting period and in extremely small amounts. Control of the concentration of the molting hormone is operated by molt-inhibiting hormones; of these, the only one identified so far is a simple alkaloid derived from tryptophan, called 3-hydroxy-L-kynurenine, the source of xanthurenic acid. Through such a mechanism, a hormone can be supplied or removed from the organism rapidly when it is no longer needed.

Zooecdysteroids were first isolated from the silkworm (*Bombyx mori*) in Europe, whereas a project in Australia, launched by the CSIRO for economical reasons, was carried out on shrimp wastes. In both cases, the ecdysteroids were

---

[8]This contradicts recent ideas that species at high latitudes are adapted to large areas as they must tolerate wide annual fluctuations in atmospheric conditions whereas tropical species are restricted to small areas as they are used to only a narrow range of changes.

151

isolated in extremely small amounts despite the large amount of arthropods used (two milligrams from two tons of crustacean residues); this posed severe problems in the elucidation of the molecular structure.

Ecdysteroids of different structures are contained in large amounts in various terrestrial plants, often over 1% of the dry weight of the plant; this situation is similar to the one encountered with the zoantharian *Gerardia savaglia* in chapter 7.

Stomatopod crustaceans, commonly called mantis shrimps, are unique in the animal world in having a kaleidoscopic vision. A typical case is *Pseudosquilla ciliata*, which inhabits all littoral tropical seas except the east Pacific. This small mantis shrimp has compound eyes with at least ten spectrally different classes of photoreceptors which result from many visual rhodopsin-type pigments and carotenoid filters; this amounts to ten different channels to distinguish different colors. In order to appreciate how extraordinary this feature is, one should consider that the most advanced multicolor vision previously known in animals is based on five different classes of photoreceptors (e.g. in some fish and butterfly retinae). Our own visual system is based on only three channels.

What is the purpose of the multichannel system in mantis shrimps? These diurnal bright-colored species are adapted to the kaleidoscopic world of the clear waters of the tropics populated by many other bright-colored species. Since the hue of many mantis shrimps varies considerably from individual to individual, multicolor vision may serve in the recognition of individuals. Mantis shrimps use the human spectral range, from blue to red, whereas multichannel-based systems of vision in most other animals extend to either the ultraviolet or the infrared region.

Perhaps the best known horseshoe crab is *Limulus polyphemus* since it served to establish modern theories of the way heartbeats are controlled. Horseshoe crabs may contain lethal toxins of the family saxitoxin-gonyautoxins of the dinoflagellates, but with a spotty distribution only. Thus, only 4 out of 388 specimens of the horseshoe crab *Carcinoscorpius rotundicauda*, collected in estuarine waters of Thailand, were found to be lethal to mice, and it is not clear whether the variability

in toxicity can be attributed to regionality or not. In view of the dietary origin of the toxins in *C. rotundicauda* and similar scavenger and detritus feeder animals, it is curious that their gills proved more toxic than their gut and intestine.

Carcinoscorpius rotundicauda

LUMINOUS SEED SHRIMPS AND KRILL

Many arthropods are bioluminescent such as sea spiders (chelicerates of the class Pycnogonida), copepods (which are the dominant marine herbivores, feeding on phytoplankton), and even large crustaceans and terrestrial milli- and centipedes. However, the phenomenon is best understood at the molecular level in shrimps and shrimplike organisms. The small seed shrimp (*Cypridina hilgendorfii*), for example, is an ostracod crustacean which commonly occurs along the coasts of Japan where it gathers on dead fish. The luminescence in *Cypridina* is an extracellular secretion which is produced when the crustacean is disturbed. From one kilogram of these crustaceans enough of the light emitting compound, 'cypridinaluciferin', has been isolated; structural investigations have assigned it to the alkaloids. And the *Cypridina* luciferase, the enzyme which induces light emission from the luciferin, has also been extracted and purified in a crystalline form.

The bioluminescence of the krill *Euphausia pacifica*, a pelagic shrimplike crustacean which forms swarms, migrates in response to light, and emits light itself, has most peculiar features. 'Euphausialuciferin' is the first known porphyrin derivative to function as a light emitter and is structurally related to chlorophylls rather than to heme. Furthermore, after that the molecular structure of 'euphausialuciferin' had been elucidated, one recognized that the dinoflagellate luciferin must be of the same type.

153

11 Bryozoans and brachiopods

The bryozoans (moss animals) and the brachiopods (lamp shells) are related by having the same type of food-taking organ, the lophophore, which lies beneath the tentacles. Together with the phoronids, these animals are therefore called lophophorates. The bryozoans are also called ectoprocts, from *ecto*, outer, and *proktos*, anus.

The bryozoans count over 4000 marine and freshwater species that compete for space and food. Though numerous and of much concern as fouling organisms in marine installations, the bryozoans are largely unknown to the non-specialist, except for some ornamental calcified species like *Sertella beaniana* or *Myriapora truncata* (false coral). Better known in North Atlantic areas is *Alcyonidium gelatinosum*, which gets caught in gill nets and is thus responsible for the dogger bank itch among fishermen. It is not surprising that we know so little about bryozoans: these sessile, encapsulated animals are inconspicuous and therefore are hard to see without a microscope.

The most primitive of all living bryozoans inhabit freshwater while stoloniferan forms, like *Zoobotrion*, are the most primitive of the bryozoans in the sea.

CYTOTOXIC SUBSTANCES OF BRYOZOANS

While nothing is known about natural products of freshwater bryozoans, marine bryozoans have been recognized as productive organisms. They secrete a glue, both at the larval and at the adult stage, which helps them to stick on every surface in the sea. This is the reason for their huge success as fouling organisms. Actually there is nothing special about this glue: it is composed of polysaccharides and proteins like the glue of other marine invertebrates, such as barnacles, mussels, and sea squirts.

BRYOZOANS AND BRACHIOPODS

As far as unusual secondary metabolites are concerned, the most productive bryozoans belong to the orders Cheilostomata (suborder Anasca) and Ctenostomata. Such bryozoans are specifically preyed upon by sea slugs to which they transfer antifeedant and toxic products for defense. Within the Anasca, the cosmopolitan fouling sea-mat *Bugula neritina* (collected in California, the Gulf of Mexico, Florida, and the Gulf of Sagami) contains complex acetogenins with powerful inhibitory action on the proliferation of lymphocytic PS leukemia. Because of their origin and

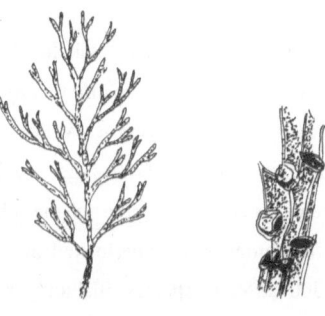

Bugula neritina

biological activity, these compounds are called bryostatins. The biological activity of the bryostatins probably stems from the propensity of the macroring of these compounds to sequester metallic ions by coordination to oxygen atoms, as "teeth", in the inside of the macroring. International health agencies regret, however, that the restricted spectrum of action of the bryostatins does not leave much hope for a practical application of these compounds.

Two other species of Anasca, *Flustra papyracea* and *Flustra foliacea* of the northeast Atlantic, contain a variety of brominated (and chlorinated) alkaloids which are not responsible, however, for the antifouling activity of global extracts of these animals.

Bryozoans of the order Ctenostomata also produce bryostatins, in particular *Amathia convoluta*. But Ctenostomata are peculiar in that *Alcyonidium gelatinosum* produces a small sulfurated compound that resembles modern synthetic reagents and is responsible for the dogger bank itch.

155

REPELLENCY OF BRACHIOPODS: A PROBLEM BEYOND THE LIMITS OF PRESENT TECHNOLOGY?

Although the brachiopods, which are exclusively marine animals, resemble bivalve mollusks from the outside, they are sessile organisms like all other lophophorates. According to fossil records, which are most abundant, the phylum Brachiopoda appeared in the Cambrian and reached a peak in the Ordovician, with 30,000 different species belonging to the class Articulata. The two calcareous halves of these brachiopods are interlocked and can only open to a certain extent. Today, less than 300 different species, including some species of the class Inarticulata are left; in the latter, the two halves, which can be made of either calcium carbonate or -unusually for shells- calcium phosphate, are held together by muscles only.

As revealed by the examination of sediments, after the Paleozoic the brachiopods became rare and the mussels abundant; this is attributable to the fact that articulates, when alone, loose space and food to the mobile mussels, as shown in experiments with living species of the Pacific coasts of Washington and British Columbia. Brachiopods have the particular disadvantage of not being able to relocate; once detached they can not reattach. But when predators, such as snails, sea stars, and crabs, are present, the result is different: mussels are predated while brachiopods are avoided. From these experiments it has been concluded that early articulates left space to the mussels and then gradually evolved deterrent compounds, either via *de novo* synthesis or by incorporation from microbial symbionts, as a protection from predators.

Unfortunately it has proven difficult to isolate the protective substances of brachiopods. We have been frustrated by this problem. In fact, our collection of (undetermined) brachiopods taken in the pre-Antarctic area of the Crozet Islands during the 1982 cruise of the French research vessel Marion Dufresne provided us with strongly fungicidal extracts but failed to afford enough organic material for study. Our large collections in Corse (by dredging with the CNRS ship Korotneff) of the articulate *Gryphus vitreus*, which is predated by annelids, also led to too little organic material to carry out purification and structural work.

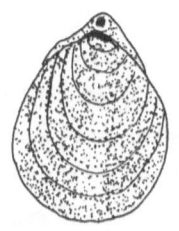
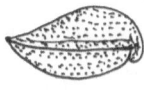

Our failure to define bioactive compounds from brachiopods may be surprising since modern analytical instrumentation is advertised to be sensitive to trace amounts of compounds. In

Gryphus vitreus

practice, however, if the nature of the compound to be monitored is not already known, about half a milligram to a few milligrams of the compound, depending on its complexity, are needed to characterize it. Such amounts of products are difficult to obtain from brachiopods, the natural products of which will therefore remain a mystery until the next revolution in analytical technology.

But our failure to solve the above problems might induce to think differently. In view of the hard work required to crash hard shells, why should predators prefer the meagre brachiopods to the fat mussels?

12 Echinoderms

Everybody knows about sea urchins and starfish. These spiny-skinned animals, the typical image of the sea in art and folklore, are part of a larger group which also comprises the holothurians (sea cucumbers), the ophiures (basket stars and brittle stars, or serpent stars), and the crinoids (sea lilies and other beautiful species which look like sessile thin basket stars). All together they constitute the phylum Echinodermata. These animals have a wide distribution on the sea bottom at all depths, but are absent from brackish waters as they lack a system for regulating osmotic pressure.

The echinoderms have a long history, revealed by fossil records comparable in abundance to those of the arthropods, the brachiopods, and the mollusks. Fossil records of echinoderms date back to the upper pre-Cambrian and are particularly abundant from the late Paleozoic, showing a general sessile condition that has been inherited by present day crinoids only. Thus, though the current form of crinoids only first appeared in the Ordovician, these animals must be considered as the most primitive of the echinoderms.

PIGMENTS OF CRINOIDS AND SEA URCHINS

The crinoids, being sessile animals, are morphologically distant from all other echinoderms. As far as their pigments are concerned, however, the crinoids are quite similar to the sea urchins. This has been recognized in crinoids of the genera *Comanthus*, *Comantheria*, *Cromatula*, and *Lamprometra*.

If there are natural products that attract widespread interest, then they

Comanthus bennetti

are the pigments: colors stimulate everyone's fantasy. And the sea is a model of vivid colors, particularly in the tropics. It is thus not surprising that since the last century sea urchins have been investigated for the pigments in their spines. All such pigments are oxidized aromatic substances that resemble the central nucleus of anthracycline antibiotics. The main pigment of the sea urchins, echinochrome-A, which inhibits both gram-negative and gram-positive marine bacteria, is so abundant and easy to extract that the process can be carried out in practical sessions by novices in the study of natural products.

The inhibitory effect of echinochrome-A in gram-positive and gram-negative bacteria is so similar that, considering the vast difference in the organization of the cell walls for the two types of bacteria, we must conclude that the cell walls are not affected by the antibiotic. This, and the slowness of the effect of echinochrome-A, which takes between one and two full days to become manifest, suggest that the antibiotic acts on bacterial metabolism.

The spectrum of activity of global extracts of sea urchins is wider: cyanobacteria are also inhibited, though the responsible agents have not been identified.

FERTILIZATION, EPA, AND GROWTH FACTORS IN SEA URCHINS: THE IMPACT ON HUMAN AFFAIRS

Inhibition of cell division of fertilized sea urchin eggs is a widely used test in the systematic search for bioactive products; the test is inexpensive, easy to carry out, and positive compounds have a good chance of being cytotoxic for abnormal cells. Since cytotoxicity tests take a long time to carry out, it is clear why the test for inhibition of cell division of fertilized sea urchin eggs has become so popular.

The mechanism of cell division of fertilized sea urchin eggs, while still obscure in detail, involves oxidation-reduction with sulfurated amino acids; this was shown by Neapolitan scientists using *Paracentrotus lividus*.

MACROFAUNA

On another front, Russian investigators have found that the sea urchin *Strongylocentrotus intermedius* is a rich source of EPA, which we already know as an antiatherosclerotic agent of diatoms.

But the most striking of all biochemical observations with echinoderms is that the sea urchin *Strongylocentrotus purpuratus* produces a polypeptide homologous to human epidermal growth factors. Such growth factors, which are definitely not restricted to mammals, having recently been found in insects as well, are responsible for cell differentiation, proliferation, and neoplastic transformations in mammals.

SIMILARITY IN CHEMICAL ARMOR OF SEA STARS, SEA CUCUMBERS, AND BRITTLE STARS

Sea stars contain hemolytic substances which are called asterosaponins for their soap-like surfactant activity. These substances deter most potential predators already at the larval planktonic stage of the starfish and act as toxins to paralyze prey. An example is acanthaglycoside-A of the starfish *Acanthaster planci* (L.) of Ehime Prefecture, Japan.

However, in the unbiased adaptational changes of Nature, the offensive compound has also become an alarm compound for potential prey. This can be seen in certain bivalve mollusks which, in response to asterosaponins released into seawater by starfish, swim away by violently expelling water from their shell. And even the sessile sea anemones have been shown to abandon the substratum and propel themselves away in response to asterosaponins. Only certain prosobranch and nudibranch mollusks prey upon starfish, which further proves that mollusks are exceptional in their ability to adapt to unconventional ambient conditions.

As expected of substances that elicit such a powerful escape response in invertebrates, the asterosaponins show a number of other biological activities including spawning inhibition, antimicrobial, antiviral and cytotoxic activities as well as other cellular activities, specifically in the follicle cell.

Brittle stars contain similar saponins, whereas sea cucumbers, which are very slow growing echinoderms, contain structurally different saponins, called holothurins; the sugars are bound to a terpene which has the skeleton of lanosterol (which in animals is the precursor of cholesterol). A case in point is

Actinopyga sp.

echinoside-A of the holothurian *Actinopyga echinites* (Jaeger) of Okinawa.

Certain terrestrial plants contain saponins, too, and their sap is used to kill and capture fish.

Starfish have specialized chemical weapons that the other echinoderms do not. The northeast Pacific sea star *Dermasterias imbricata* contains an alkaloid, imbricatine, which induces in *Stomphia coccinea* the same swimming response as asterosaponins do in sea anemones.

GROWTH AND DEATH OF CORAL REEFS: THE *Acanthaster* PHENOMENON

Sweet in abstract painting, ferocious in actual behavior, crawling about slowly on the sea bottom with its armor of powerful physical and chemical weapons: this is the

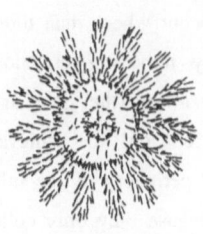

Acanthaster planci

starfish. A particularly aggressive species is the crown-of-thorn starfish, *Acanthaster planci* (L.), which has recently multiplied at a high rate in southern tropical Pacific areas. Because it feeds exclusively on scleractinians, it is bringing about the destruction of large portions of coral reefs. The starfish feeds by extruding its stomach against the coral and digesting it. In a few hours the stomach is retracted leaving a round dead patch the size and form of the starfish.

At the Great Barrier Reef the first recorded outbreak of *Acanthaster* occurred in 1964 around Green Island, a beautiful island covered by trees and situated in front of Cairns; this was followed by another outbreak in 1979 in the same area and today there are still no signs of regression. Millions of starfish can be counted there, whereas the normal population density of starfish is six per square kilometer. Now the starfish larvae transported by the currents and the adults moving in search of food are shifting the *Acanthaster* phenomenon toward the south.

In Australia there is much concern about the *Acanthaster* phenomenon as Japanese and Australian developers hesitate in investing in tourist resources along the Great Barrier, as well as in other infested areas such as Fiji and the Tonga Islands, for fear of the *Acanthaster* infestation spreading.

Opinions about the causes of the outbreak and methods of controlling it are contradictory. Certain Australian scientists have noted that this is not a phenomenon of our industrial society and that severe infestation by *Acanthaster* had already occurred when the Great Barrier was untouched. These scientists do maintain that collecting *Acanthaster* in the infested areas, as already attempted in Okinawa, will not control the outbreak. They think the only way to control the phenomenon is to better understand coral reef areas, in particular the life cycle of *Acanthaster*.

Other Australian scientist, however, negate the validity of these conclusions. They have observed that outbreaks of *Acanthaster* always occur where man threatens the reef by overfishing, or unbalances the coral reef by removing predators of *Acanthaster*, such as the prosobranch mollusc *Charonia tritonis* (the giant triton). They believe that, in the short term, control of *Acanthaster* can only be achieved by removing all specimens, including the small ones, from the periphery of the infested area. They maintain that the Japanese failed in Okinawa because they only collected the large specimens of starfish. In the long run, the only valid way to control

ECHINODERMS

Acanthaster, according to these scientists, is to return to the traditional methods of fishing and to leave coral reef areas undisturbed. This, I believe, is also the best way to prevent other as yet unforeseen ecological disasters on coral reefs.

13 Ascidians

The ascidians (sea squirts) belong to the chordates which are the most evolved of the invertebrates. The chordates are animals equipped with a hollow dorsal nerve cord, at least at some stage of their development.

It is thus useful to separate the class of the sessile sea squirts, together with two much smaller classes of planktonic species, the most common and best known of which are the salps, in the subphylum lower chordates; these are alternatively called urochordates or tunicates. The term "lower chordates" implies lower evolution, "urochordates" means that the animals have a tail, and "tunicates" is derived from the solid tunic with which these animals are covered. The tissue of the tunic of the sea squirts is unusual for animals because it is made of a polysaccharide (called tunicin) whose molecular structure is similar to the cellulose of plants.

The sea squirts are solitary or colonial filter-feeding organisms of worldwide distribution; they are subdivided into three orders: the Phlebobranchia, the Stolidobranchia, and the Aplousobranchia. Solitary ascidians have two openings to the exterior: the mouth, into which water is pumped, and a smaller excurrent aperture from which water is pumped out. Colonial ascidians have maintained independent mouth openings, but excurrent openings converge in a common cloacal cavity.

SEA SQUIRTS AS SPECIALIZED SINKS FOR METALS IN UNUSUAL STATES

Ever since the time of Aristotle, the Phlebobranchia have been the most familiar of the sea squirts. The philosopher was fascinated by these animals, probably by *Ciona intestinalis*, the most common species in the Mediterranean. It is a long tubular transparent species which grows everywhere on submerged objects, including boats. Along the coasts of my native Tuscany, these ascidians are called "pinci" and the saying "mettere i pinci" (to let sea squirts grow on oneself) is used to describe people who do not like to move.

ASCIDIANS

Interest in sea squirts was kindled anew at the beginning of our century, when it was discovered that certain blood cells of the Phlebobranchia have the ability to selectively concentrate metals. The type of metal preferentially sequestered, mostly vanadium, iron, molybdenum, and niobium, is species specific.

Recently it has become apparent that a warm water species of Phlebobranchia, *Ascidia nigra* (L.), which is covered by a solid black tunic, is capable of concentrating vanadium by one million-fold in a reduced form that is normally unstable in water media under physiological conditions. Vanadium can be stored in such an unusual form because of a reducing and complexing bright yellow pigment called tunichrome-B1, which is found in active blood cells called vanadocytes. This pigment is a small peptide derived from an oxidized form of the amino acid DOPA (which is a building block of our skin and hair polymeric pigments, the melanins). Tunichrome-B1 is so sensitive to both water and molecular oxygen that its extraction from ascidians proved quite troublesome; it took years to isolate 0.5 milligrams of pure compound from 4000 ascidian specimens.

Exciting conjectures have been made as to the role of the vanadocytes and of related cells, the ferrocytes, of other sea squirts. Such cell types seem to be involved in the immune system of the sea squirts and one wonders whether tunichrome-B1 may be the evolutionary precursor of compounds with similar functions in man.

PHARMACOLOGY WITH APLOUSOBRANCH SEA SQUIRTS

Though the title "Fighting cancer and AIDS with sea squirts" is not yet warranted for this section, we are struck by a remarkably high incidence of cytotoxic and antiviral compounds in sea squirts of the order Aplousobranchia. The genera *Trididemnum, Didemnum,* and *Lissoclinum* (family Didemnidae) which often form large plates in the tropics, typically contain special cyclic peptides of the depsipeptide class. Thus, *Trididemnum solidum* (Van Name) of the Caribbean, as well as other didemnides, contain didemnin-B which enhances antibody production and is currently in Phase II of clinical examination as an anticancer agent. This compound is scarcely available in Nature, i.e. only in trace amounts in the didemnides. Therefore, much effort went into the total synthesis of this compound in various chemical laboratories. Thanks to such projects didemnin-B is now available in substantial amounts.

Certain didemnides contain peptides: for instance the peptide ulicyclamide has been isolated from *Lissoclinum patella* of the Western Caroline Islands. This work was carried out in Honolulu and in Hawaiian "uli" means a dark color, as is the case with this dark green ascidian.

Other antineoplastic compounds are present in didemnides. Two species of Okinawa, *Didemnum* sp. and *Cystodytes dellechiajei*, contain powerfully antileukemic alkaloids, ascididemnin and cystodytin-A, respectively. Ascididemnin is a representative of an alkaloidal class that is found not only in other ascidians of the family Didemnidae, such as *Leptoclinides* sp., but also in sponges, such as *Amphimedon* sp. (Haplosclerida) and *Petrosia* sp. (Petrosiida), and sea anemones, such as *Calliactis parasitica*.

Didemnone-C is a small, structurally different, antileukemic acetogenin of the didemnides *Didemnum voeltzkowi* of Suva Harbor, Fiji, and *Trididemnum* cf. *cyanophorum* of Shroud Cay, Bahamas.

Cytotoxic compounds are also present in colonial aplousobranchs of the family Polyclinidae. In *Aplidium* *californicum* (sea pork), collected in the intertidal zone near the Golden Gate Bridge, the responsible compound is prenylquinone, whereas with an orange-flecked *Aplidium* sp. of the Gulf of California the active compound is a terpenoidic aminoalcohol called aplidiasphingosine.

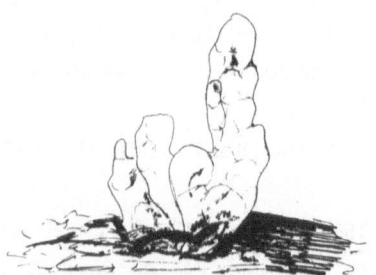

Aplidium californicum

ASCIDIANS

Another collection of *Aplidium californicum* in an estuary near Bahia Kino in Pacific Mexico proved to contain cytotoxic macrolides called bryostatins, which had already been isolated from certain bryozoans. It was in fact proven that the bryostatins of this sea pork originate from the bryozoan *Bugula neritina*, an interstitial species in the ascidian colony.

Colonial Aplousobranchia of the family Polycitoridae contain bioactive compounds with good prospects as antiviral agents. *Eudistoma olivaceum*, collected among the roots of mangroves in Atlantic Mexico, Banco Chinchorro, or in Belize, at Lighthouse Reef, contain various alkaloids, called eudistomins, which are strongly active against *Herpes simplex* in monkey kidney tissues and show low cytotoxicity.

Structurally similar alkaloids, called eudistomidins, which exert calmodulin antagonism, have been extracted from the green colored ascidian *Eudistoma glaucus*, collected at a depth of 5-10 meters near Ie Island, Okinawa.

COMPARATIVE BIOCHEMISTRY AS A SURROGATE FOR FOSSIL RECORDS

Being made of soft tissues, like contemporary species, the early tunicates have left no fossil records. Their history can only be inferred after examining modern species and is therefore poorly known.

A complementary approach to the history of these animals was taken in my laboratory, in collaboration with French scientists. We comparatively examined the distribution of natural products in urochordates and other phyla and were able to show that colonial didemnides, starfish, and acorn worms contain unusual nucleosides like 2'-deoxyuridine, isolated from *Trididemnum cereum* (Giard, 1872). This supports a phylogenetic relationship among urochordates, hemichordates, and echinoderms, as suggested on classical biological grounds.

167

MACROFAUNA

14 Fishes

OF FISH AND WORMS

According to the Frankfurter Allgemeine Zeitung, 3 February 1988, Soviet researchers have found out that fish fat discarded during the preparation of fish product can be purified and used as a valuable shampoo. Patents are pending in many countries and an experimental factory is now producing 100 tons of the product annually.

While waiting for such a pearl of a shampoo, and still using the trivial synthetic shampoo of the supermarket round the corner from my house, I got another piece of information out of the press, i.e. that the fishing industry of northern Europe is experiencing a deep crisis. The decay of the fishing industry followed a journalist's scoop that the fish in northern Europe were infested with worms. The journalist failed to mention that worms came before fish according to accepted chronologies and that fish, as soon as they came, were inhabited by worms. The journalist failed also to mention that worms are not only found in fish. Pork meat may contain something that turns into a worm if ingested; for this reason we cook or roast pork meat longer than other meats. I do not know if this was merely ignorance on the journalist's part or if the worms in pork meat have been saved for the next scoop, when the time is ripe to throw the pork industry into a crisis.

Having already dealt with the luminous processes of fish and knowing little about the gelatinous mucus with which the parrotfish covers itself before sleeping in the coral reef at night, perhaps to prevent its scent from spreading around, there is not much more to be said about fishes within the framework of this book. This follows a general trend: as one moves toward more evolved organisms, the richness in natural products of the primitive organisms is lost. Such a trend in natural product distribution among the various phyla may reflect the defensive organization of all living species. The vertebrates possess an inducible immunitary defensive system, and, with few exceptions, rely on it alone, while the invertebrates have a more rudimentary non-inducible system and therefore need protection by natural products. In these conclusions I have omitted physical strategies of defense which are more or less used by all living things.

NON-IMMUNE DEFENSE OF FISH

The ciguatera poisoning is a phenomenon that we have discussed in connection with the dinoflagellates. It might be considered as a defensive system against human intrusion: along Indo-Pacific coasts where ciguatera is endemic there are only a few fishing boats and if we consider the ecological disasters brought about by heavy fishing in coral reef areas, we can conclude that the ciguatera's strategy has worked well.

Another chemical defense of certain tropical fish, in particular the puffer fish, is a potent neurotoxin called tetrodotoxin, and which we already know as a bacterial product. This toxin is structurally similar to the paralytic shellfish poisons of the dinoflagellates (saxitoxin and gonyautoxins) and acts by blocking the sodium channels in the membranes of cells by the same mechanism; it has therefore become an important tool for physiological studies.

Parrotfish liver poisoning is another chemical defense mechanism, the symptoms of which resemble those of both paralytic shellfish poisoning and puffer fish poisoning. Incidents with humans have occurred because the parrotfish *Ypsiscarus ovifrons*, though not tasty, has a liver which is considered a delicacy in southern Japan and in Indo-West Pacific areas. The powerful toxins tetrodotoxin and palytoxin, already known from other phyla, as discussed above, are responsible for these poisonings.

Trunkfish (Ostraciidae) of Australian coasts, particularly *Aracana ornata*, *Ostracion cubicus*, and *Rhinesomas reipublicae*, are toxic owing to large concentrations of nitrogenated acetogenins called choline esters.

MACROFAUNA

A special defensive system against marine predators has been set up by tropical flat fish of the sole family. We are not alone in considering the soles a delicacy, and this has posed acute problems of survival to these flat fishes in tropical areas. The problem was solved with defensive compounds. The Red Sea Moses sole, *Pardachirus marmoratus*, repels sharks when it is about to be attacked by emitting toxic peptides called pardaxins from glands located at the base of the spines of the dorsal fin. These toxins have also the properties of detergents spreading on and adhering to the pharyngeal cavity and/or the gills of the sharks.

Pardachirus pavoninus, the small peacock sole of the West Pacific, is a master in camouflage, by adapting its color to that of the surroundings and by fluffing up sand when settling on the sea bottom. But this was not a sufficient defense against predators; in Okinawa and in other tropical areas the sole had to set up a chemical defense system to keep away sharks. The active agents are pardaxins similar to those of *P. marmoratus*.

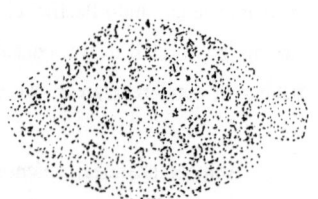

Pardachirus pavoninus
(peacock sole)

The sequence of amino acids of pardaxin-1 has been investigated and, curiously, its 33 amino acids are in a sequence similar to that of melittin, the toxic peptide of the honey bee (*Apis mellifera*).

Both soles mentioned above also produce ichthyotoxic steroidal aminosugars, mosesins and pavoninins, which also contribute to shark repellency. However, if you are wondering about a cream based on these compounds that you could use when you go swimming in Shark's Bay, you will be disappointed by the strong hemolytic properties typical of saponins that prohibit the use of these compounds for humans.

Finally, there are the active poisonous fishes, such as the scorpion fish, in which the venoms are peptides.

BELOW AND ABOVE FISHES ON THE EVOLUTIONARY SCALE

All higher fishes are jawed fishes which first appeared in the Silurian. But fish had appeared on the Earth before. In the Ordovician jawless fish were abundant; they form a link to jawed fish, having separated from the parasitic hagfishes and the lampreys, which are limbless. The secondary metabolism of jawless fish is totally obscure, however.

Amphibians come before jawed fish on the evolutionary scale; they first appeared in the Devonian and have adapted to terrestrial life. They contain a variety of natural products, including tetrodotoxin, which, as we already know, is typical of certain jawed marine fishes but is first produced by bacteria.

Moving up the evolutionary scale we meet the reptiles, whose most important period was the Mesozoic. Fossil remains of large marine reptiles, such as the pliosaur *Liopleurodon macromerus* which had a length of about 7 m, were found in Kimmeridge Clay of Westbury, Wiltshire, UK, and are now on display in the Bristol Museum. Large predators such as the pliosaurs and the ichthyosaurs, equipped with long and strong jaws with powerful teeth, inhabited the same waters at the same time. Today the oceans would not be able to support such large carnivorous species. In fact, whales have adapted as plankton feeders. Perhaps this is the most striking difference between the ecological organization of modern seas (where large, long-lived predators feed on small, short-lived plankton) and that of terrestrial habitats (where small, short-lived predators feed on large, long-lived plants).

Turtles are reptiles, and a contemporary member, the spongivorous hawksbill turtle, *Eretmochelys imbricata*, has already been dealt with in chapter 6. Green turtles are the protagonists of an heroic migration: at the end of December, every year, they migrate 2000 km from the coasts of Brasil to their nesting grounds on Ascension Island in the middle of Atlantic; this takes about two months. The reason for this long migration route is a mystery. Perhaps the choice of Ascension Island was made when green turtles begun to develop, some 40 million years ago; at that time the island was close to the coasts of Brasil and then it gradually moved further away as a result of continental drift. If so, the voyage of the green turtle from Brasil to Ascension Island is natal homing. This is suggested by the dissimilarity of genetic material of green turtles of Ascension Island from those using other beaches, off Venezuela and Florida. However, such differences in genetic material are so slight that, within the hypothesis of natal homing, sporadic changes in nesting grounds must have occurred during the long period considered.

Marine reptiles also comprise sea snakes, and I know a lot about them. Hoping for a relaxing day without sharks, I once thought it a good idea to swim in a pond on the Ilot Maître in the lagoon in front of Noumea. There were in fact no sharks, but sea snakes in unusual abundance.

Sea snakes, as air breathing animals, have to come to the surface at intervals, though certain species can stay underwater for as long as one hour. They inhabit coastal waters from northeast Africa to Okinawa, all the area at lower latitudes, Australia, the tropical south Pacific, and the Pacific coasts of central America. The most diffuse species is *Enhydrina schistosa*, particularly in Asiatic areas, and it is the more aggressive species, too. Its venom is made up of polypeptidic neurotoxins the amino acid sequence of which has been established. When they bite, sea snakes eject most of their venom; if the venom glands are full, the victim dies rapidly. This is probably the result of evolutionary change in the toxin's structure: unlike terrestrial snakes, marine snakes can not rely on the scent of the victim who moves away while dying.

But Nature is on our side this time: sea snakes have such small fangs that man can not be easily bitten, except in thin parts, such as between the fingers.

On the evolutionary scale the mammals come after the reptiles. It was in a more temperate and drier climate, which favored the formation of extensive prairies at the expense of forests, that mammals first appeared. While certain mammals became adapted to flying, like the pterosaurs, some of the mammals returned to the sea, becoming porpoises and whales, and later seals. This animal world is too evolved in the direction of the immune system to offer much to the table of the secondary metabolites, however. Scents may be an exception, but they are usually made up of small molecules, uninteresting to the natural product chemist.

APPENDICES

A. Biological fundamentals

EUKARYOTIC VS. PROKARYOTIC CELLS

Though phytoplankton is small, the cells of the smallest diatoms and flagellates can be easily seen under an ordinary microscope. However, in order to observe planktonic free-living bacteria, which barely attain a diameter of one thousandth of a millimeter, special optical methods are needed.

In all cases, to appreciate the fine details of the cells, one needs an electron microscope, which has a much higher resolving power than an optical microscope. The cells can be divided into two groups: the prokaryotic and the eukaryotic cells. Bacteria and cyanobacteria are prokaryotes, whereas all other organisms, with the exception of the archaebacteria to which we have alluded in chapter 4, are eukaryotes. In all cases biochemical studies have shown that cells are 90% water and that different functions are performed in the different organelles of the cell.

In the following an idealized eukaryotic cell will serve to illustrate the main cellular features. The

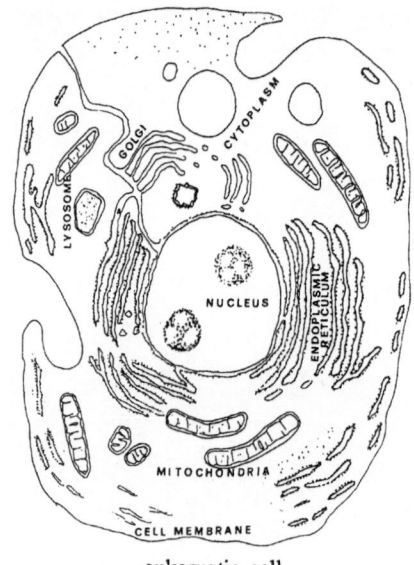

eukaryotic cell

174

cytoplasm is enclosed in a double-layered membrane, which is so thin that it appears as a single line in our drawing. In the central portion of the cell there is the nucleus surrounded by a membrane. The nucleoplasm contains a nucleolus and more than one linear chromosome, which is an association of DNA with a protein.

The cytoplasm also contains various organelles, which are membrane-enclosed spaces, such as the lysosomes, where hydrolysis of polysaccharides occurs yielding monomeric sugars. These are taken up by the mitochondria, as centers of energy metabolism, and are oxidized to carbon dioxide. Another organelle is the endoplasmic reticulum, which contains the ribosomes where the proteins are synthesized. The proteins are then packed and distributed by the Golgi apparatus.

The cytoplasm of photosynthetic cells also contains plastids or chloroplasts where photosynthesis occurs. Plant cells also have cavities, called vacuoles, for food storage. Finally, many unicellular organisms possess centrioles, small cavities near the nucleus which act as contractile or locomotory elements in the muscle function of higher animals.

As with primary metabolites, though probably with less specificity, the synthesis of secondary metabolites of different classes occurs in different compartments of the cell. For instance, certain hormonal regulators of plant growth, the gibberellins and abscisic acid, which are terpenes, are mainly produced in the plastids. Why has evolution led to this choice? The reason is probably that the plastids are particularly rich in the reactants required for the biosynthesis of these hormones.

Specialized cells are capable of accumulating secondary metabolites, such as brominated acetogenins in the red seaweed *Bonnemaisonia nootkana* and, as already noted in chapter 6, amino acid metabolites in the sponge *Aplysina fistularis*.

All the cellular characteristics of eukaryotes are absent in prokaryotes, and the genetic organization is different. In prokaryotic cells the major part of the DNA forms a single circular chromosome, not restricted within a nuclear membrane, and genetic information is also contained in extrachromosomal DNA. Some metabolic functions are also performed in a different way.

APPENDICES

While there is general consensus that plastids and mitochondria of eukaryotic cells originate from prokaryotic cells along an endosymbiont line, the opinions differ in regard to the cell nucleus: some scientists advocate an endosymbiontic origin of the nucleus as the first step in the evolution of eukaryotes, whereas other scientists have proposed that the nucleus derives from cell compartmentation.

CELL TYPE AND EVOLUTIONARY MARKS

An important point is that while the eukaryotic cell seeks to maintain its characteristics, the prokaryotic cell lacks such a strict restraint; the prokaryotic cell is able to exchange genetic material, though this occurs less frequently than is generally thought. Genetic engineers have emphasized this property of eukaryotic cells to justify their efforts in mixing the characteristics of all organisms, including eukaryotes, which were devised not to do so; or at least not at the speed of genetic manipulation. In any event, an overall improvement on Nature through genetic manipulation is yet to be seen.

This preamble serves to emphasize the fact that patterns of sequences of amino acids in proteins are related to the evolutionary history of species. As this is similar for evolutionarily-related species, such patterns can be used as evolutionary marks. An example in chapter 14 concerns the green turtles of Ascension Island. If, however, exchange of genetic material has occurred recently in prokaryotic cells, where it is allowed to happen, such patterns are altered and may not be used as evolutionary marks. Therefore, the prokaryotic cell, though more amenable to investigation than the eukaryotic cell, may be unsuited to unravel the evolutionary history of species.

PHOTOSYNTHESIS IN THE SEA

Photosynthesis occurs in the sea with or without the evolution of oxygen. When oxygen evolves, photosynthesis occurs the same way as in terrestrial plants, except for the attenuation and change in color of light with increasing depth, which influences both the distribution of marine organisms and the need for accessory pigments.

Although all animals depend on photosynthesis for survival, as they feed on products of photosynthetic organisms, the way this occurs in certain marine animals is peculiar. A case in point is that of reef-building corals and other organisms which live in a symbiosis with microscopic photosynthetic algae called zooxanthellae. The corals do not feed on zooxanthellae products in the usual sense; transfer of photosynthetic nutrients from zooxanthellae to corals occurs directly. As a consequence, reef corals need much light to support their zooxanthellae, so that shadowing seaweeds are not allowed to grow excessively.

The photosynthetic process of green seaweeds and higher plants involves decomposition of water to give intermediate compounds which have reducing properties. These intermediates react with carbon dioxide to afford sugars in a dark process. Amino acids, fatty acids, and nucleotides, which are essential compounds for the growth of all organisms, are produced in subsequent steps which are difficult to distinguish from photochemical acts. This is the primary production; solar energy and a few simple inorganic compounds, in a process mediated by a complex system, yield the organic compounds which are needed in the construction of the complex molecules of life.

It is impressive that nearly half of the total world primary production occurs in the sea although the biomass of photosynthetic organisms is far smaller in the sea than on the land. Even more striking is that net production of oxygen is higher in the sea than on the land; this is due to the fact that many marine organisms (such as reef-building corals and algae, foraminifera, mollusks, calcareous sponges, and

ascidians belonging to the family Didemnidae) have
the ability to fix carbon dioxide yielding calcium
carbonate. This process counteracts the
overproduction of carbon dioxide which results from
massive fossil fuel burning on the land. One should
therefore pay more attention to the conservation of
the sea resources and not be only concerned with the
problems of the Amazonian forest which is no net
oxygen producer.

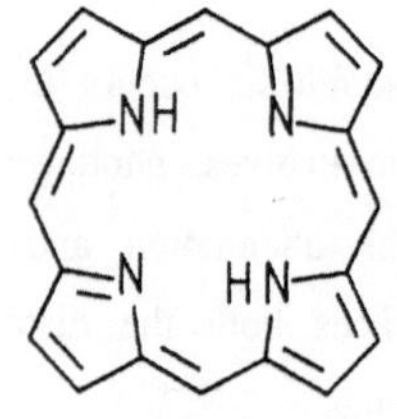

porphin

All organisms capable of photosynthesis with oxygen evolution use
chlorophyll-a to induce splitting of the water molecule, so that chlorophyll-a has

been taken as an index of
phytoplankton biomass. This has found
practical application with the advent of
satellites, which have replaced the
slow-cruising research ships. Satellites
furnish real-time maps of the
distribution of chlorophyll-a which
serve in the remote sensing of oceanic
primary production through appropriate
algorithms. Green algae and higher
plants also have chlorophyll-b,
whereas brown seaweeds, diatoms, and
dinoflagellates have in its place
chlorophyll-c, and red seaweeds

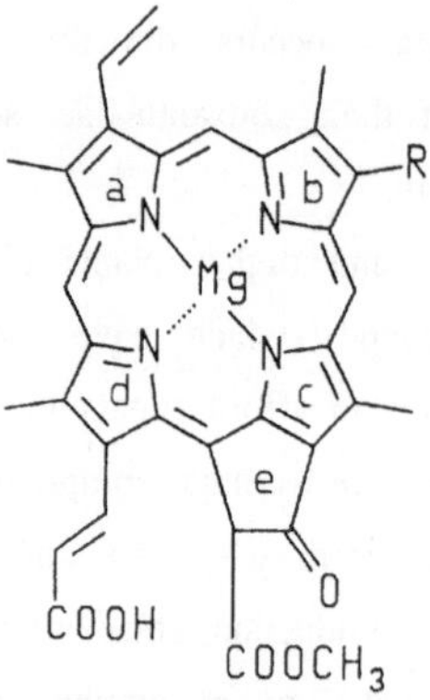

chlorophyll-c
(mixture of compounds with
$R = CH=CH_2$ or CH_2CH_3)

have chlorophyll-d.

All chlorophylls, in particular chlorophyll-c, are structurally related to heme,
the prosthetic group of hemoglobin which is the most common biological carrier of

oxygen[9]. While heme contains iron, all chlorophylls contain magnesium and chlorophyll-c (actually a mixture of two compounds) has the tetrapyrrole structure of

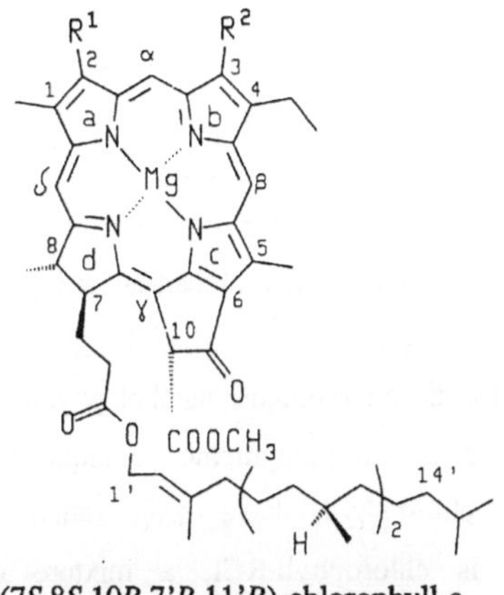

(7S,8S,10R,7'R,11'R)-chlorophyll-a
(R^1 = CH=CH$_2$, R^2 = CH$_3$)
(7S,8S,10R,7'R,11'R)chlorophyll-b
(R^1 = CH=CH$_2$, R^2 = CHO)

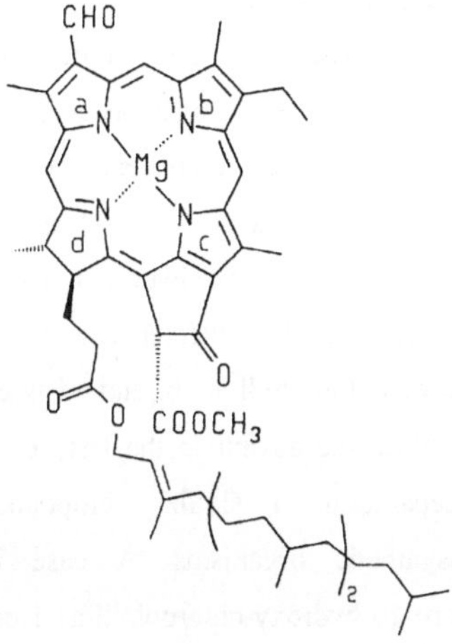

chlorophyll-d

heme, with different substituents, and the closure of ring e, at the hypothetical "porphin" nucleus. The magnesium ion is at the center of the tetrapyrrole unit, bound to the four nitrogen atoms; the same occurs in the other chlorophylls, where, however, reduction of the porphin nucleus has occurred. Derivatives of chlorophyll-c, where magnesium has been exchanged for nickel or vanadium, are widely distributed

---

[9]Non-heme oxygen-transporting conjugated proteins in which metal atoms are coordinatively bound to amino acids are also known. In hemerythrin, the oxygen carrier of echiurid and sipunculid worms, as well as of brachiopods, there are two Fe(II) per place of coordination of O$_2$. On binding oxygen by hemerythrin as peroxidianion (O$_2^{2-}$), Fe(II) changes to Fe(III) with a corresponding change of color from pale yellow to red. In hemocyanin, the oxygen carrier of many arthropods and mollusks, there are two Cu(I) per place of coordination of O$_2$. As a result of oxygen binding by hemocyanin, there is a change from Cu(I) to Cu(II) with a corresponding change from colorless to blue.

in oil shales of the Cretaceous period. Quite recently another porphyrin-type chlorophyll, chlorophyll-c3, has been isolated from the brown alga *Emiliania huxley* (Prymnesiophyceae).

Porphyrins are also represented by iron-coordinated prosthetic groups of respiratory pigments such as myoglobin and cytochromes, as well as by the prosthetic groups of certain enzymes, such as catalases and peroxidases.

Chlorophyll-a, -b, and -d have the structure of modified porphyrins, called chlorins, in which two hydrogen atoms are added to the external carbon atoms of ring d. As a further difference with respect to chlorophyll-c, the carboxylic group at ring d of chlorophyll-a, -b, and -d is esterified by the diterpenoidic alcohol phytol.

With the advent in the last decade of powerful chromatographic techniques for the separation of similar compounds, other chlorophylls have been found in photosynthetic organisms. A case in point is chlorophyll-RCI, a mixture of δ-chloro-10-hydroxy-chlorophyll-a isomers which has widespread occurrence in oxygenic photosynthetic organisms.

CARBON VS. SILICON AS A BASIS OF LIFE

$$O=C=O$$

carbon dioxide

The photosynthetic process, on which life depends, begins with carbon dioxide, which is found both in the atmosphere and dissolved in the waters and can be taken up by photosynthetic cells and added of active hydrogen to give sugars and other key compounds. Sugars burn with oxygen in living organisms to give back carbon dioxide and water, which can be used again in the photosynthetic process.

Fictional science, in the search for the unusual, has imagined an extraterrestrial life based on silicon, on the grounds that silicon is the element chemically closest to carbon. However, the resemblance of silicon dioxide, SiO_2, to carbon dioxide, CO_2, in terms of the same relative proportion of atoms, is purely formal. Whereas single CO_2 molecules do exist, SiO_2 molecules do not. The linear arrangement of the CO_2 molecule allows the two oxygen atoms to stay as far away from each other as possible, in order to minimize repulsions. In contrast, in silicon dioxide there is a continuous network arrangement of oxygen atoms bridging two silicon atoms, which results from the joining of tetrahedra with silicon at the center.

Silicon-oxygen double bonds are not formed as they are weaker than silicon-oxygen single bonds, which is just the contrary of what occurs with carbon and oxygen. This deprives silicon dioxide of the facile reduction-oxidation capability of carbon. In conclusion, carbon is unique in that it is able to change from the inorganic state of carbon dioxide to the organic state of the molecules of life accompanied by the production of molecular oxygen, and then to return to the inorganic state when life molecules burn with oxygen in living organisms. In the case of silicon, such processes are unfavorable, and do not occur. This fact, as well as its inability to form chains and rings, makes silicon an unsuitable backbone of life molecules.

THE ACCESSORY PIGMENTS OF PHOTOSYNTHESIS: A PHYLOGENETIC MARK FOR RED SEAWEEDS

In addition to light-harvesting chlorophyll-b, or in its stead, algae and cyanobacteria possess other pigments which have the same function and are therefore called accessory pigments. While the most general of these pigments are the carotenoids, cyanobacteria and red seaweeds also possess biliverdin-type pigments: phycoerythrobilin in red seaweeds and phycocyanobilin in cyanobacteria. These

181

pigments, which are bound to a protein as phycobiliproteins in Nature, attest the

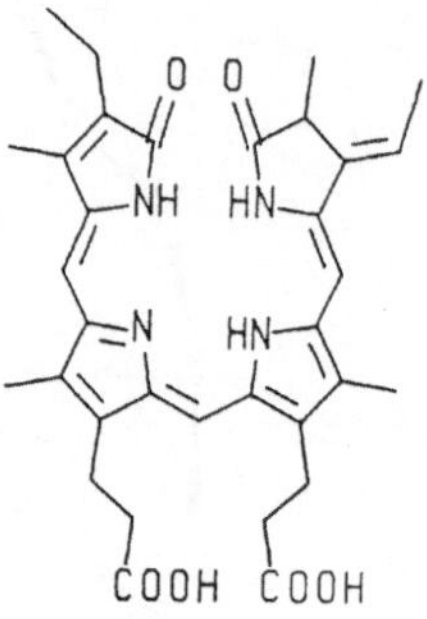

phycocyanobilin 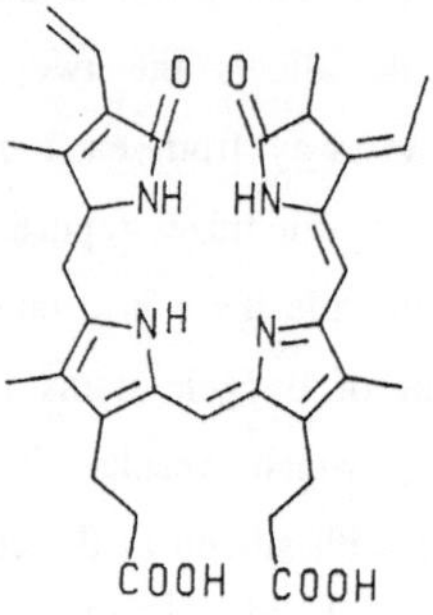   phycoerythrobilin

origin of the red seaweeds from cyanobacteria. Phycoerythrobilin and phycocyanobilin take, as free molecules, the helical form depicted here; when they are bound to a protein, the helical form is lost in favor of an elongated form.

BACTERIOCHLOROPHYLLS: THE NON-OXYGENIC PHOTOSYNTHESIS

In their detoxification processes, phototrophic bacteria use bacteriochlorophylls instead of the chlorophylls of photosynthetic eukaryotes and cyanobacteria. Purple non-sulfur bacteria, such as *Rhodopseudomonas*, carry out photosynthesis using bacteriochlorophyll-a and -b, whereas green and brown sulfur bacteria, such as *Chlorobium* and *Chloropseudomonas*, use bacteriochlorophyll-c, -d, and -e which are also called *Chlorobium* chlorophylls.

In all cases the bacteriochlorophylls are coordinated to magnesium in the same way as the chlorophylls; bacteriochlorophylls b-e carry two extra hydrogen atoms at ring d and therefore are dihydroporphyrins while bacteriochlorophyll-a carries two more hydrogen atoms at ring b and therefore is a tetrahydroporphyrin.

An analog of bacteriochlorophyll-a with *all-trans*-geranylgeraniol instead of

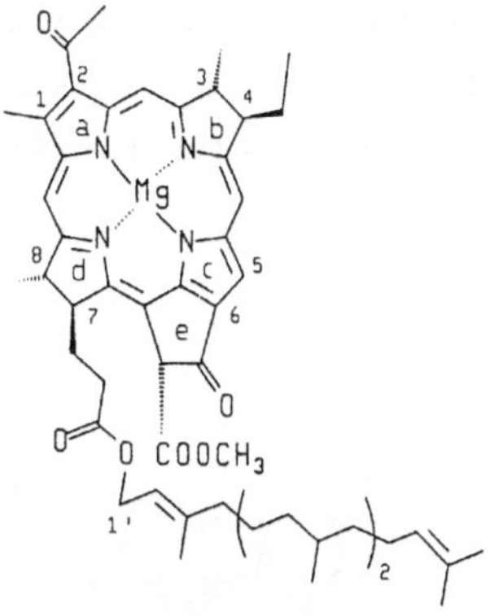

(3R,4R,7S,8S,10R)-
bacteriochlorophyll-a

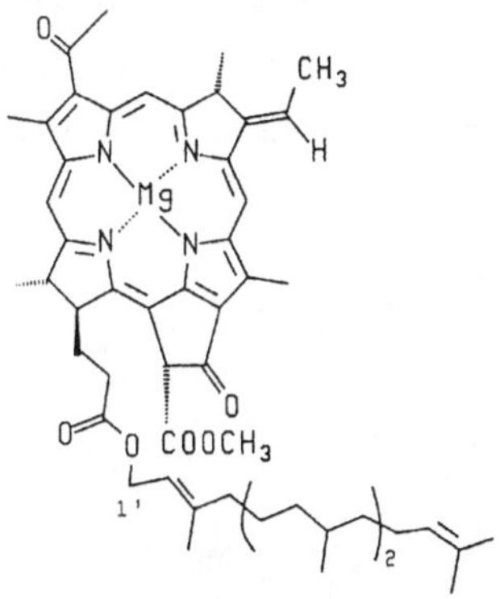

(3R,7S,8S)-
bacteriochlorophyll-b

phytol (i.e. with C(6')=C(7') and C(10')=C(11') *trans* double bonds) has been found in the purple non-sulfur bacterium *Rhodospirillum rubrum*, and other bacteriochlorophylls are also known or will be discovered with the aid of modern, powerful chromatographic techniques.

Bacteriochlorophyll-c is a mixture of compounds. It is also referred to as *Chlorobium* chlorophyll-660 since the absorption in the visible region occurs at the wavelength of about 660 x 10⁻⁹ meters. Actually, six absorption bands can be detected in such a region; four of these bands pertain to the compounds with R^1 = Bu^i or = Et, whereas the other two bands are each for a mixture of two

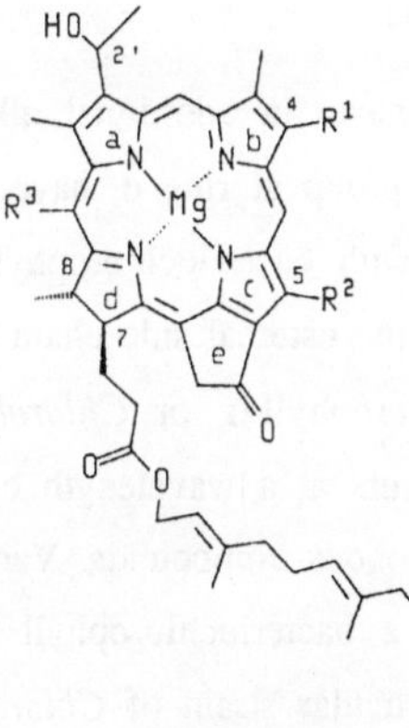

	R^1	R^2	R^3
	Bu^i	Et	Et(a
	Bu^i	Et	Me(b
	Pr^n	Et	Et(c
	Pr^n	Et	Me(c
	Et	Et	Me(b
	Et	Me	Me(b

(7S,8S)-bacteriochlorophyll-c
a) = (2'S)
b) = (2'R)
c) = (2'R) and (2'S)

183

diasteromers which differ from one another in the configuration at C(2'). One couple of these diastereomers has R^1 = Pr^n and R^2 = R^3 = Et and the other couple has R^1 = Pr^n, R^2 = Et, and R^3 = Me. Thus, with respect to bacteriochlorophyll-a

	R^1	R^2
	Bu^i	$Et^{(a}$
	Pr^n	$Et^{(b}$
	Bu^i	$Me^{(a}$
	Et	$Et^{(c}$
	Pr^n	$Me^{(c}$
	Et	Me
	Np	$Et^{(a}$
	Np	$Me^{(a}$

	R
	$Et^{(a}$
	$Pr^n{(a}$
	$Bu^i{(a}$

(7S,8S)-bacteriochlorophyll-d
a) = (2'S)
b) = (2'R) or (2'R) and (2'S)
c) = (2'R)

(7S,8S)-bacteriochlorophyll-e
(a = (2'R) and (2'S)

these molecules have an additional alkyl group at the δ carbon atom[10], lack the methoxycarbonyl group at ring e, have an alcoholic function at the chain of ring a, and, in analogy with bacteriochlorophyll-d and -e, have the C_{15} farnesol instead of the C_{20} phytol at the estereal side chain of ring d.

Bacteriochlorophyll-d, or *Chlorobium* chlorophyll-650 (the absorption in the visible region occurs at a wavelength of about 650 x 10^{-9} meters), also represents a mixture of homologous compounds. Various absorptions bands can be detected, each corresponding to a bacteriochlorophyll with the R^1/R^2 substituents specified in our drawing; in a particular strain of *Chlorobium*, however, the case R^1 = Pr^n and R^2 = Et corresponds to two diastereomers differing from one another in the configuration at C(2').

---

[10]Abbreviations for the R alkyl substituents are as follows: Me = -CH_3, Et = -CH_2CH_3, Pr^n = -CH_2CH_2CH_3, Bu^i = -CH_2CH(CH_3)_2, and Np = -CH_2(CH_3)_3.

Bacteriochlorophyll-e is a mixture of six homologous compounds (in each of the three couples of diastereomers the difference lies in the configuration at C(2')).

A more extensively hydrogenated porphyrin-like tetrapyrrole compound (factor F430), where the four nitrogen atoms are coordinated to nickel, are used by methanogenic bacteria in carrying out the reduction of carbon dioxide to methane.

THE NATURE OF THE PHOTOSYNTHETIC PROCESSES

In purple photosynthetic bacteria e.g., which as prokaryotes are easier to study than eukaryotes, a membrane-bound sandwich-like pair of bacteriochlorophyll molecules receives a unitary amount of light whereby an electron is transferred to a nearby molecule of a pigment called bacteriopheophytin. Since the electron bears a negative charge, the place from which it is taken remains positively charged whereas its place of relocation becomes negatively charged. At the membrane of the photosynthetic apparatus such a charge separation is stabilized by an extremely rapid (2 x 10^{-10} seconds) transfer of an electron to a ubiquinone molecule.

In other words, the primary act of photosynthesis is a conversion of absorbed photons into an electrical voltage difference at the membrane. A fine point is that the molecules involved in these electron transfers are separated from one another by 10 Å, which is a large distance, considering the fact that bonding electrons reside between nuclei which are only about 1 Å far apart in molecules. How can an electron transfer take place over such a large distance? It does so because of the bizarre nature of this particle: the better its energy is determined, the worse can its position be described. In other words, the electron can go through the energy barrier for relocation, rather than having to go over it, as the thickness of the barrier is comparable to the uncertainty of the position of the electron in the space. Such a process is appropriately called tunneling.

Perhaps the very reason for the extreme efficiency of the photosynthetic process, which is the fastest known photochemical reaction between two molecules, has not yet been fully grasped, however. Recent studies with the isolated reaction centers of the purple photosynthetic bacteria *Rhodopseudomonas viridis* and *Rhodobacter sphaeroides* have in fact made the matter even more intriguing than heretofore thought. In such studies crystallographic tridimensional determinations of the structures of the reaction centers were carried out at the resolution of interatomic distances while studies of the reaction centers with ultrafast visible spectroscopy allowed the detection of the initial species, a charge-transfer complex of molecules.

$$2 \underset{H}{\overset{}{A}}{-}H \quad + \quad O{=}C{=}O \quad \longrightarrow \quad \underset{H}{\overset{H}{C}}{=}O \quad + \quad H_2O \quad + \quad 2A$$

generalized photosynthetic process

The primary chemical act which follows charge transfer is the splitting of a compound of type H_2A whereby nicotinamide adenine dinucleotide phosphate (NADP) is reduced to NADPH; the latter is the actual reducing agent of carbon dioxide to formaldehyde (H_2CO). At the same time adenosine diphosphate (ADP) is changed to adenosine triphosphate (ATP), which is the phosphorylating agent. Overall the process may be depicted as where A represents a generic atom (such as oxygen, in which case $H_2A = H_2O$) or a group of atoms, or even nihil (in which case $H_2A = H_2$, as with certain bacteria). In other words, the electron flow in the photosynthetic process leads to the oxidation of H_2A and reduction of CO_2. While higher plants and algae obtain electrons from water, photosynthetic bacteria can get electrons from a variety of other molecules, such as hydrogen sulfide (H_2S) or other oxidizable sulfur compounds, molecular hydrogen (H_2), or even small organic compounds made by other organisms.

Being able to imitate the conditions of voltage difference in the photosynthetic process is man's dream in devising new solar batteries for energy conversion and setting up a photosynthetic reactor to produce sugars and other primary products.

Significant advances in this direction have recently been made, though photosynthesis remains one of the most difficult processes to imitate because of the fine-tuned, close, cooperative action of so many partners.

SCREENING OF NATURAL PRODUCTS FOR BIOLOGICAL ACTION

Currently there are five main lines of interest in marine natural products. They concern the ecological role, the control of fouling and other noxious organisms, the pharmacological action, the exploitation of marine natural products as tools in physiology as well as developmental organic chemistry.

In assessing the role exerted by marine natural products in controlling marine organisms, the choice is between actual life conditions (which are most difficult to set up, as I have mentioned in chapter 7) and laboratory conditions. What has not yet been taken seriously into account is the control of subtly noxious humans. I agree with Schopenauer that there are many humans who are far more dangerous than the white shark or the scolopendra of the tropics.

Some of the problems involved with the screening of marine natural products for pharmacological action have already been discussed with the diatoms. In general, a drug, to exert useful pharmacological effects, has to fulfill many requirements that can not be tested *in vitro*. The first requirement is that translocation of the drug to the site of action occurs efficiently, i.e. that the pharmacokinetics is favorable. If this requirement is met, the drug has to be non-toxic at the dose required for pharmacological action.

Other practical points concerning biological tests have already been dealt with, in particular the fact that preliminary indication of cytotoxicity can be obtained quickly from tests for the inhibition of the cell division of fertilized sea urchin or starfish eggs (chapter 12).

BIOLOGICAL CLASSIFICATION

The first modern system of biological classification appeared in the year 1735 with the first edition of the "Systema naturae", written by Carolus Linnaeus. The Swedish naturalist, ranked to nobility in 1761 as Carl von Linné, and subsequently the anatomists George Cuvier in France and Richard Owen in Great Britain, established relationships among living species on the basis of similarities. Linnaeus' binomial system (whereby a living, or extinct, organism is indicated by two Latin names, the first one for the genus and the second one for the species) is still in use. Genera which bear similarities are grouped into families, families into orders, orders into classes, and these into phyla or divisions. Currently, in describing a particular organism, the following descending hierarchical order is adopted

superkingdom
   kingdom
      subkingdom
         phylum (or division)
           class
              subclass
                order
                   suborder
                     family
                        genus
                           species
                             variety

This classification system has many problems, first of all in the definition of a species as a group of interbreeding organisms; in captivity different species belonging to the same genus may breed with each other. Even the definition of a genus as a set of related species is problematic since establishing similarities is largely a subjective judgment. Many other criteria have recently been introduced, or resumed from the past, in taxonomy; they are based on embryological observations (as we have seen for the sponges), computer analysis of behavioral traits or all traits,

BIOLOGICAL FUNDAMENTALS

and observations that no two species can have exactly the same habitat in the same area.

But perhaps the most unbiased criterion of biological classification has Darwin as its father. After the theory of evolution was introduced, taxonomic studies became part of the studies on the evolution of species in the sense that the closer two species are to a common ancestral organism, the closer the two species are placed in the taxonomic scale. In our times such relationships are best based on similarities in proteins (as we have seen in chapter 14 for the green turtles of Ascension Island) or in genetic material (as we have seen in chapter 4 for bacteria). To what extent secondary metabolites may serve the same purpose is an intriguing matter which I have deferred to appendix C.

In conclusion, taxonomy, being continuously faced with so many challenging problems, has many practical applications, and requires the scientist to have a profound knowledge in many areas. Changes in the way of classifying living organisms follow the progress in our knowledge and no complete agreement on any one system of classification has ever been reached. In this book I have adopted a compromise system of biological classification that should satisfy the needs of my readers; the list below is largely restricted to taxa of marine interest and is detailed down to class or subclass only.

superkingdom Prokaryotae (formerly Protokaryotes, from *protos*, first,
and *karyon*, kernel)
kingdom Virus (from *virus*, poison)
kingdom Monera
division Bacteria
division Cyanophyta (from *kyanos*, heavy blue, and
phyton, plant)
division Prochlorophyta (or Protochlorophyta, from *protos*, first, *khloros*,
green, and *phyton*, plant)
superkingdom Eukaryotae (from *eu*, true; *karyon*, kernel)
kingdom Plantae (plants)
subkingdom Thallobionta (lower plants)
division Rhodophyta (from *rhodon*, rose, and *phyton*, plant; red algae)

189

division Chromophyta (from *khroma*, brown, and *phyton*, plant; brown
 algae)
    class Bacillariophyceae (diatoms)
    class Chryptophyceae (cryptophytes)
    class Dinophyceae (dinoflagellates)
    class Phaeophyceae (brown seaweeds)
division Euglenophyta
division Chlorophyta (from *khloros*, green, and *phyton*, plant; green
 algae)
    class Chlorophyceae (green seaweeds)
division Eumycota (fungi)
subkingdom Embryobionta (higher plants)
    division Magnoliophyta
    class Monocotyledonae
    order Najadales (Seagrasses)
kingdom Animalia (animals)
subkingdom Protozoa (from *protos*, first, and *zoa*, animal)
subkingdom Parazoa
    phylum Porifera (sponges)
    class Calcarea (calcareous sponges)
    class Demospongiae (spongin sponges)
        subclasses Homoscleromorpha and Ceractinomorpha
 (viviparous spongin sponges)
        subclass Tetractinomorpha (oviparous spongin sponges)
    class Hexactinellida (glass sponges)
    class Sclerospongiae (possibly a polyphyletic group, not a class)
subkingdom Eumetazoa
    phylum Cnidaria (from *cnida*, nettle)
    class Anthozoa (from *anthos*, flower, and *zoa*, animal)
        subclass Alcyonaria (= Octocorallia)
        subclass Zoantharia (= Hexacorallia)
    class Cubozoa (true medusae)
    class Hydrozoa (from *hydra*, a multi-headed monster, and *zoa*,
 animal; hydroids)
    class Scyphozoa (from *scyphos*, cup, and *zoa*, animal)
    phylum Ctenophora (from *ctenos*, comb, and *phoros*, to bear)
    phylum Platyhelminthes (from *platy*, flat, and *helminthos*, worm)
    class Cestoda (tapeworm)
    class Trematoda (flukes)
    class Turbellaria (free-living flatworms)
    phylum Nemertea (proboscis or ribbon worms)
    phylum Annelida
    class Oligochaeta (from *oligo*, few, and *chaeta*, bristle)
    class Polychaeta (from *poly*, many, and *chaeta*, bristle)

BIOLOGICAL FUNDAMENTALS

phylum Nematoda (roundworms)
phylum Echiuria (spoon worms)
phylum Sipunculida (peanut worms)
phylum Arthropoda (from *arthros*, joint, and *pod*, leg)
    class Crustacea (from *crusta*, shell)
phylum Chelicerata (king crabs, sea spiders, mites)
phylum Mollusca (mollusks, from *molluscus*, soft)
    class Gastropoda (from *gaster*, belly, and *podos*, foot)
        subclass Opisthobranchia (from *opistho*, back, and *branchia*, gill)
        subclass Prosobranchia (from *proso*, in front, and *branchia*, gill)
        subclass Pulmonata (from *pulmonis*, lung)
    class Cephalopoda (from *kephale*, head)
    class Bivalvia (bivalves)
    class Scaphopoda (from *scaphe*, boat, and *podos*, foot; elephant's tusk shells)
phylum Phoronida (horseshoe worms)
phylum Bryozoa (moss animals, from *bryon*, moss and *zoa*, animal)
phylum Brachiopoda (from *brachium*, arm, and *pod*, foot; lamp shells)
phylum Chaetognata (from *chaetae*, curved spikes, and *gnata*, jaws; arrow worms)
phylum Echinodermata (from *echinos*, spiny, and *derma*, skin)
    subphylum Asterozoa
        class Stelleroidea
            subclass Asteroidea (from *aster*, star, and *zoa*, animal; sea stars)
            subclass Ophiuroidea (from *ophis*, snake; brittle stars and basket stars)
    subphylum Crinozoa (from *krinon*, lily, and *zoa*, animal; crinoids)
    subphylum Echinozoa
        class Echinoidea (sea urchins, from *echinos*, spiny)
        class Holothuroidea (from *holothurion*, a water polyp; sea cucumbers)
phylum Hemichordata (from *hemi*, half, and *chorda*, rod; acorn worms)
phylum Chordata (from *chorda*, rod)
    subphylum Tunicata (tunicates, from their tunic) (= Urochordata, from *oura*, tail)
        class Ascidiacea (from *asc*; sea squirts)
        class Thaliacea (from *Thaleia*, one of the three Graces)
    subphylum Cephalochordata (from *cephalon*, head, and *chorda*, cord)
    subphylum Vertebrata

B. Classification and structure of the natural products discussed above[11])

B1A. INORGANIC PROCESSES

process

occurrence/role

$NH_3 \rightarrow NH_2OH \rightarrow NO_2^- \rightarrow NO_3^-$
ammonia hydroxylamine nitrite nitrate

nitrification carried out
by gram-negative bacteria
(ammonia is the main
base in the sea); it makes
nitrogen available to
plants.

$8 \, S^{2-} \rightarrow S_8 \rightarrow SO_3^{2-} \rightarrow SO_4^{2-}$
sulfide sulfur sulfite sulfate

sulfur utilization by
gram-negative bacteria as
energy source.

B1B. INORGANIC COMPOUNDS

structure

isolated
from/bioactivity/role

I^-

seawater (0.0005 g/l) and
seaweeds. It is
incorporated in the
thyroid (hormone
tyroxin): an iodine
deficiency leads to goiter
in humans; incorporated
also in the endostyle of
cephalochordates of the
genus *Branchiostoma*.

---

[11])Classification of the metabolites here is generally based on likely biogenesis and
not on biosynthetic evidence. Mixed biogenesis is not accounted for in the
classification. No absolute configuration is implied by the following structural
formulae, unless absolute-configuration notation or chiroptical data are reported. Trivial
names coined by this author for natural products are enclosed within single quotation
marks.

NATURAL PRODUCT STRUCTURES

I₂
Br⁻ and Br₂

seaweeds; antibacterial.
seawater and seaweeds;
biosynthetic pathways to
bromocompounds are
stimulated by
peroxidases.

Cl⁻

seawater; responsible, as
sodium salt, for the high
osmotic pressure of
seawater.

B2. CARBOHYDRATES

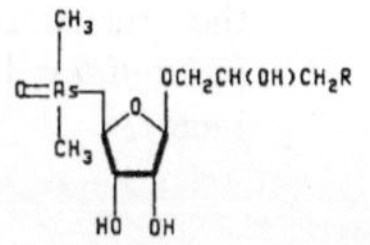

'eckloniarsenoribofuranoside' (R = SO₃H)
'tridacnarsenoribofuranoside' (R = OSO₃H)

the brown seaweed
Ecklonia radiata
(Laminariales) and the
bivalve mollusc *Tridacna
maxima*; poison.

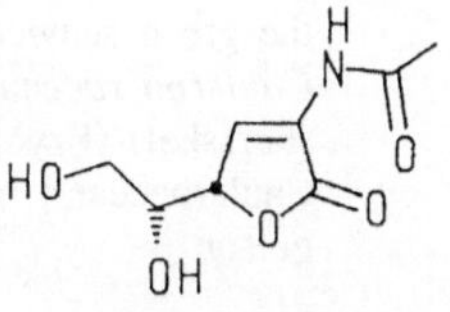

(+)-leptosphaerin

cultures of the
ascomycete *Leptosphaeria
oraemaris* (Linder).

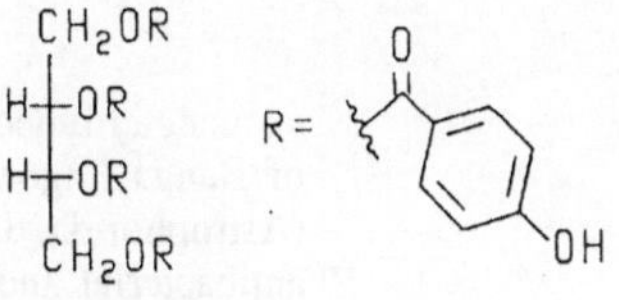

kelletinin-I

the tropical prosobranch
mollusc *Kelletia kelletii*;
related metabolites from
the Mediterranean
prosobranch *Buccinulum
corneum*; antibacterial.

193

B3. ACETOGENINS
B3.1. FATTY ACIDS, THEIR DERIVATIVES, AND HYDROCARBONS
B3.1A. FATTY ACIDS

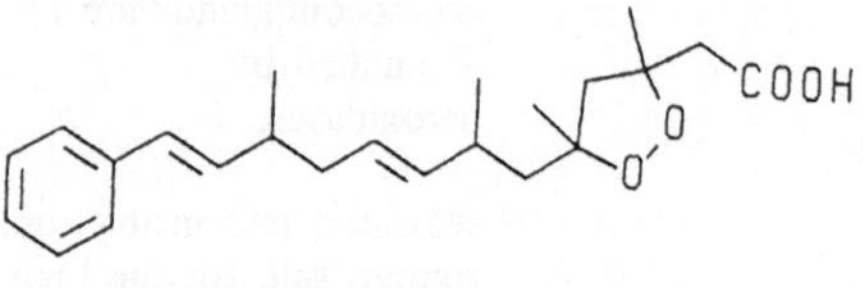

plakinic acid-A

an undetermined sponge
of the Caribbean
(Homoscleromorpha,
Plakinidae); antifungal.

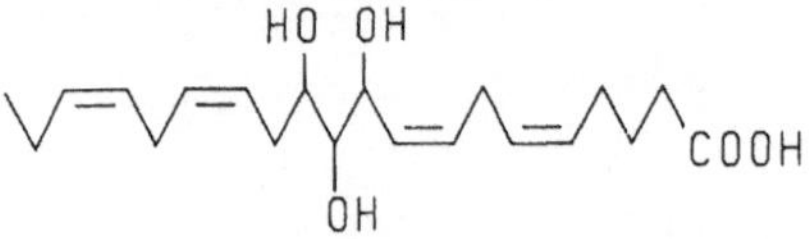

'balanoacid'

the crustacean *Balanus
balanoides*; hatching
factor.

B3.1B. FATTY ACID AMIDES, NITRILES, AND ARSENODERIVATIVES

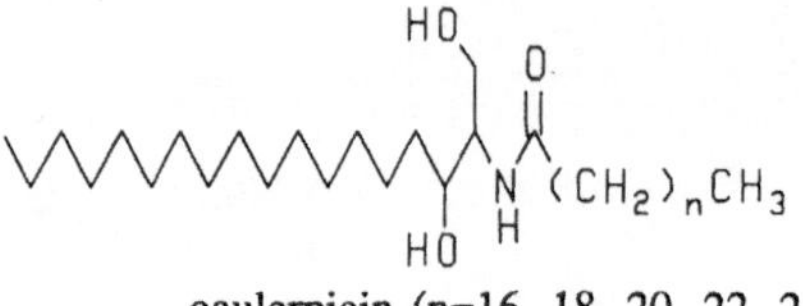

caulerpicin (n=16, 18, 20, 22, 24)

the green seaweed
Caulerpa racemosa
(Forsskal) (Bryopsidiales,
Caulerpaceae); mild
poison.

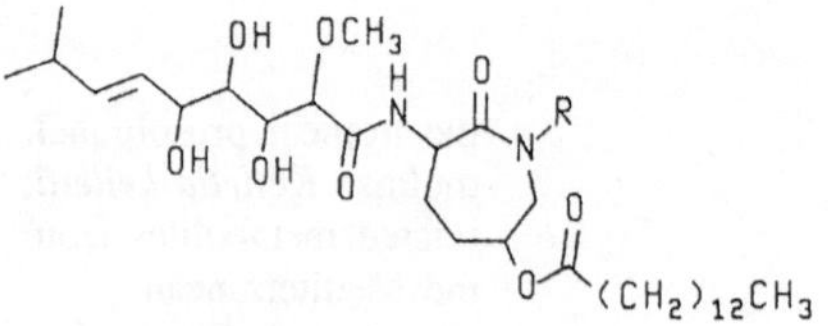

bengamide-B (R = CH₃)

an undetermined sponge
of Benga Lagoon, Fiji
(Astrophorida, Jaspidae);
antibacterial and
anthelmintic.

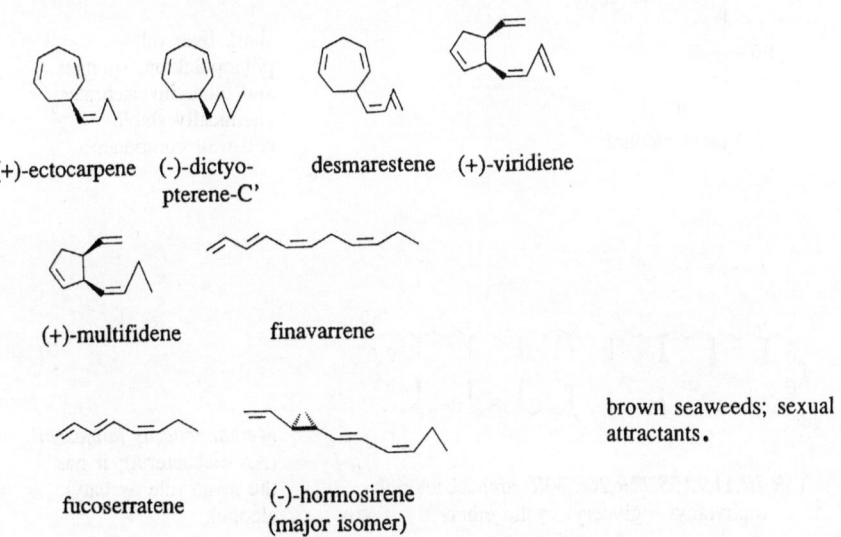

calyculin-A

the sponge *Discodermia calyx* (Spirophorida); antileukemic.

(CH₃)₃As⁺CH₂COO-
arsenobetaine

the shrimp *Sergestes lucens* and various sharks and fish; poison.

B3.1C. HYDROCARBONS AND HALOCARBONS

(+)-ectocarpene   (-)-dictyo-   desmarestene   (+)-viridiene
pterene-C'

(+)-multifidene   finavarrene

brown seaweeds; sexual attractants.

fucoserratene   (-)-hormosirene
(major isomer)

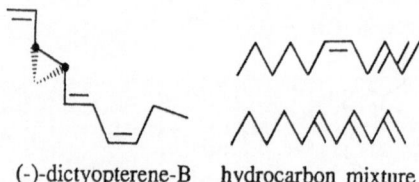

(-)-dictyopterene-B   hydrocarbon mixture

brown seaweeds of the genus *Dictyopteris* (Dictyotales); responsible for the "ocean smell".

CH_3Br $CH_3CH_2CH_2Br$
bromomethane 1-bromopropane

various red and brown seaweeds; pollutants.

B3.1D. POLYACETYLENES

duryne

the sponge *Cribrochalina dura* (Haplosclerida); antitumoral.

B3.2. GLYCERYL ETHERS

$$\begin{array}{l} {-}O(CH_2)_{17}CH_3 \\ HO{-}H \\ {-}OH \end{array}$$

(+)-batyl alcohol

shark liver oil, phytoplankton, sponges, and other invertebrates; chemically stable cell-wall components.

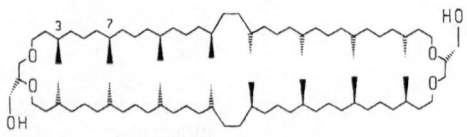

(3*R*,7*R*,11*R*,15*S*,22*R*,26*R*,30*R*)-archaebacterial diphytanyl diglyceryl cyclic ether

Methanococcus jannaschii (Archaebacteria); it has the same role as batyl alcohol.

B3.3. POLYETHERS

brevetoxin-B
CD(MeOH) Δε = -3.93(225), +6.77(257)

cultures of the dinoflagellate *Gymnodinium breve* [= *Ptychodiscus brevis*] (Davis) (Gymnodiniales); it affects the sodium channels at the membranes and causes repetitive firing in neurons; inhibitor of the cell division of fertilized sea urchin eggs.

ciguatoxin

moray eel (*Gymnothorax javanicus*); a congener toxin from *Gambierdiscus toxicus* (Dinoflagellates); broad spectrum of symptoms in man, including skin irritation and respiratory failure.

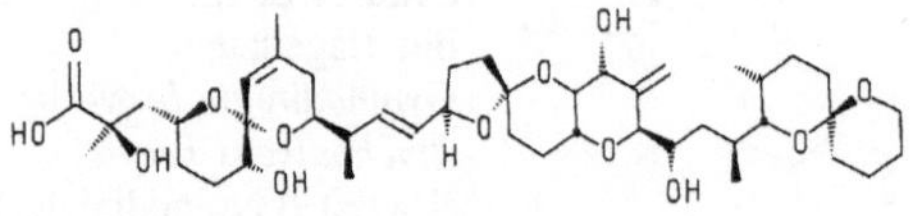

(+)-okadaic acid

cultures of the dinoflagellate *Prorocentrum lima* (Prorocentrales); the sponges *Halichondria* [= *Reniera*] *okadai* (Kadota) and *Halichondria melanodocia* (Halichondrida), and mussels; antitumoral.

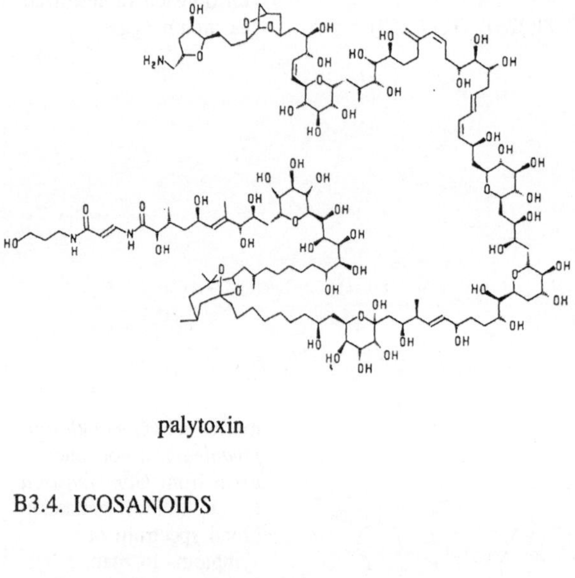

palytoxin

bacteria; the zoanthids *Palythoa toxica, Palythoa tuberculosa*, and *Palythoa caribeorum*; the crab *Demania reynaudii*; the parrotfish *Ypsiscarus ovifrons*; it induces bradycardia, rapid breathing, cyanosis, and block of renal functions.

B3.4. ICOSANOIDS

arachidonic acid

the venom duct of the neogastropod mollusc *Conus textile*; precursor of bioactive metabolites (prostaglandins, punaglandins, thromboxanes, leukotrienes).

198

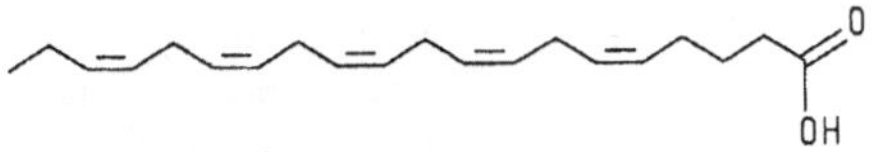

EPA

the diatom *Phaeodactylum tricornutum*; fish oil; antithrombic and antiatherosclerotic.

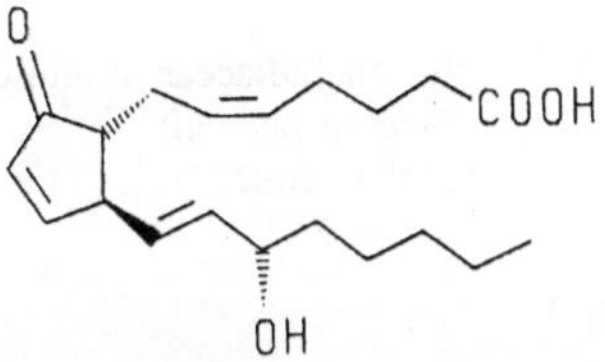

(15S)-PGA₂

the semen of mammals; the gorgonian *Plexaura homomalla* (Esper) forma *kükenthali* Moser (as methyl ester) (*Plexaura homomalla* (Esper), forma *homomalla*, contains (15R)-PGA₂) it induces a lowering of the blood pressure and muscular contraction; inflammatory agent.

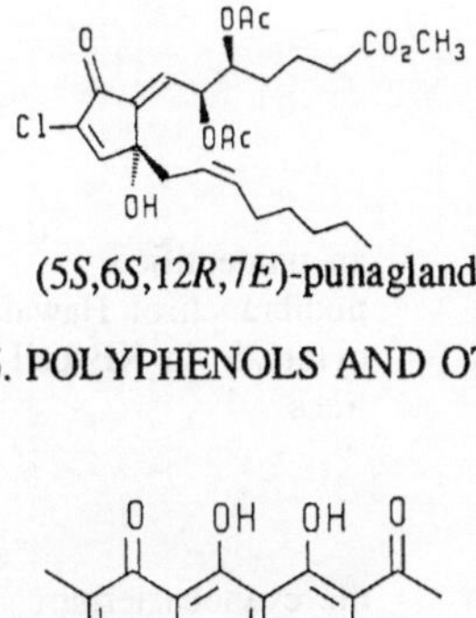

(5S,6S,12R,7E)-punaglandin-4

the telestacean octocoral *Telesto riisei*; cytotoxic.

B3.5. POLYPHENOLS AND OTHER CYCLIZED POLYKETIDES

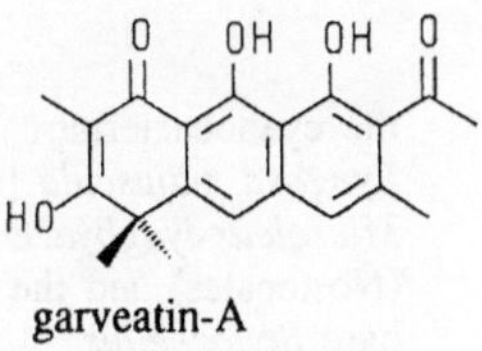

garveatin-A

the hydroid *Garveja annulata*; antimicrobial.

B3.6. PYRONES

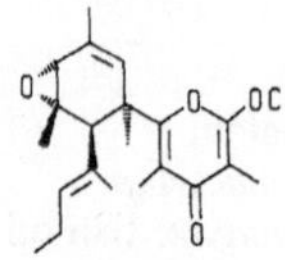

tridachione

the sacoglossan opisthobranch *Placobranchus ocellatus* (Van Hasselt); it acts as a protective filter from solar radiation.

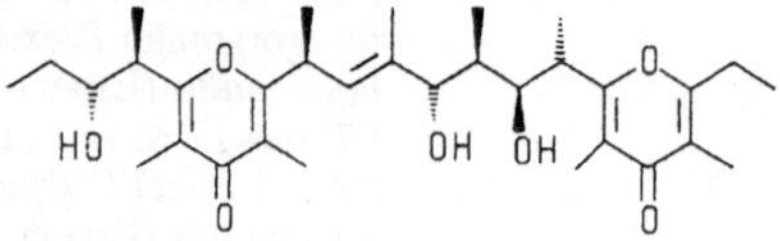

peroniatriol-I

the onchidiacean mollusc *Peronia peronii*; antileukemic.

B3.7. MACROLIDES

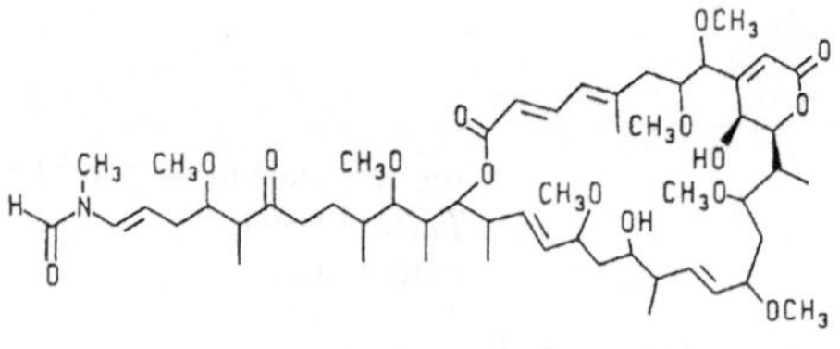

sphinxolide

an unidentified nudibranch of Hawai; cytotoxic in KB cell lines.

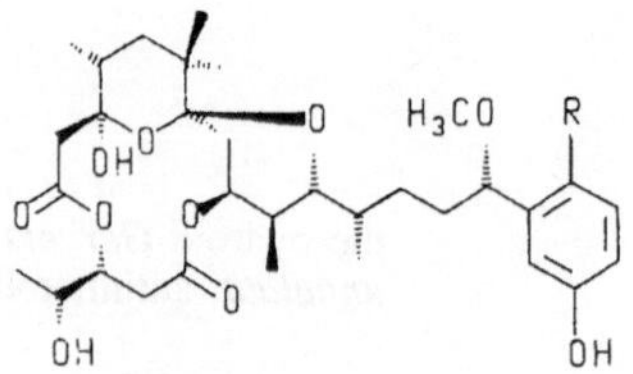

aplysiatoxin (R = Br)
debromoaplysiatoxin (R = H)

the cyanobacterium *Lyngbya majuscula* [= *Microleus lyngbyaceus*] (Nostocales) and the sea hare *Stylocheilus longicauda*; responsible for contact dermatitis; antifeedant.

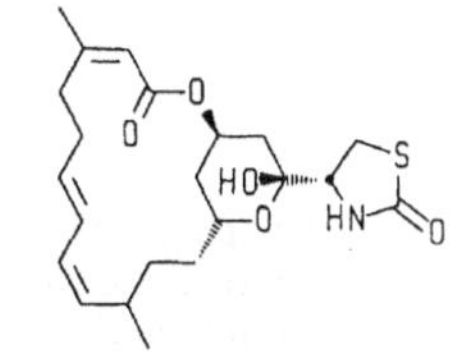

(+)-latrunculin-A

the sponge *Latrunculia magnifica* (Keller) (Hadromerida) of the Red Sea and the nudibranch *Chromodoris elisabethina* of Guam and Eniwetok; ichtyotoxic (by bringing about disruption of microfilament cellular organization).

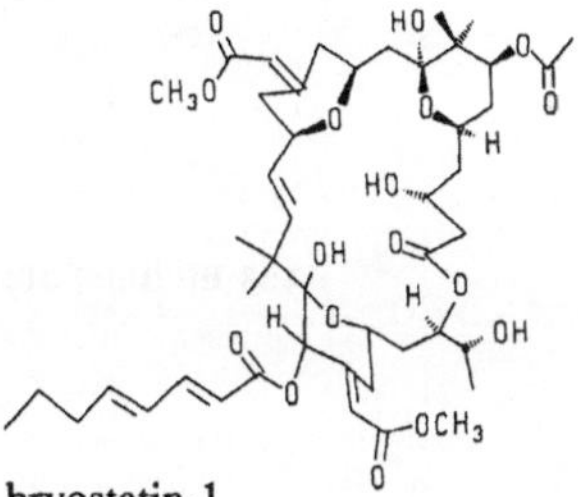

bryostatin-1

the bryozoans *Bugula neritina* (L.) (Cheilostomata) and *Amathia convoluta* (Anasca); antileukemic.

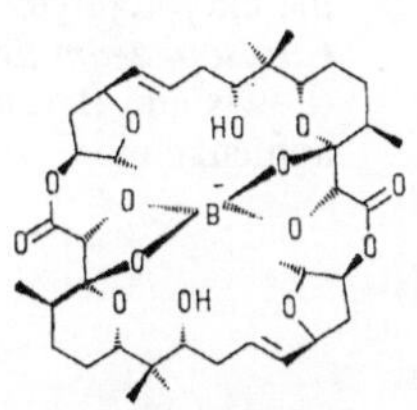

(+)-aplasmomycin

cultures of the bacterium *Streptomyces griseus*; inhibitor of *Plasmodium* and bacteria.

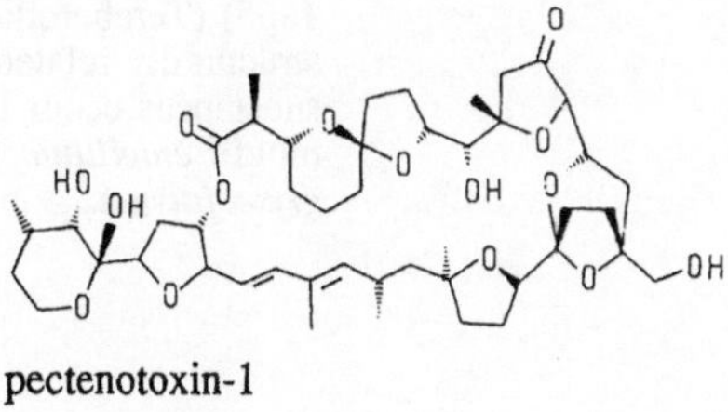

pectenotoxin-1

scallops and mussels of Japanese waters, though of dinoflagellate origin; diarrhetic shellfish poison.

201

B4. SHIKIMATES
B4.1. QUINONES AND POLYPHENOLS

xestoquinone

the sponge *Xestospongia sapra* (Petrosiida); cardiotonic.

echinochrome-A

sea urchins; antibacterial.

hormothamnione

the chryptophyte *Chrysophaeum taylori* (Lewis and Bryan); antileukemic.

B4.2. OTHER SHIKIMATES

thelepin

the annelid *Thelepus setosus* (Quatrefages, 1865) (Tereberellidae); structurally related substances occur in the mold *Penicillium griseofulvum*.

B5. TERPENES
B5.1 HEMITERPENES

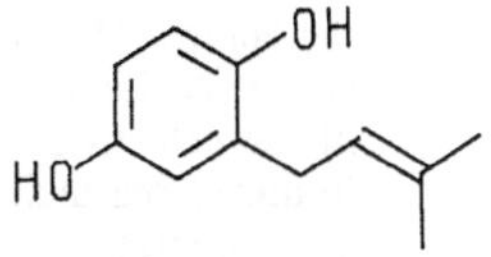

the seasquirt *Aplidium californicum* (Polyclinidae); antitumoral *in vivo* (leukemia in mice).

prenylquinone

B5.2. MONOTERPENES

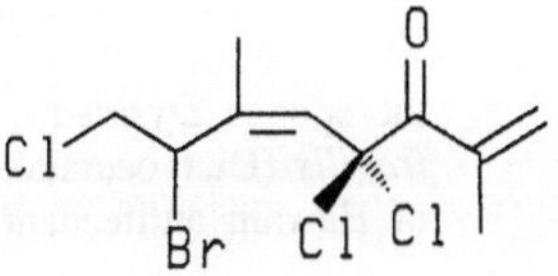

the tropical red seaweed *Plocamium* sp. (Gigartinales); odoriferous.

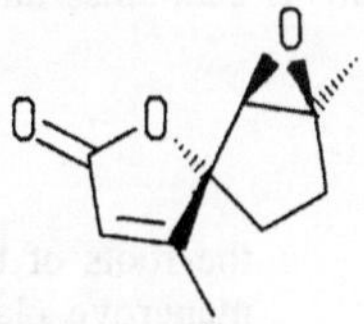

plocamenone

the sponge *Dysidea* sp. of Venice (Dictyoceratida).

adriadysiolide

B5.3. SESQUITERPENES
B5.3A. Derived via 6-11 cyclization of a farnesyl precursor: the monocyclofarnesanes

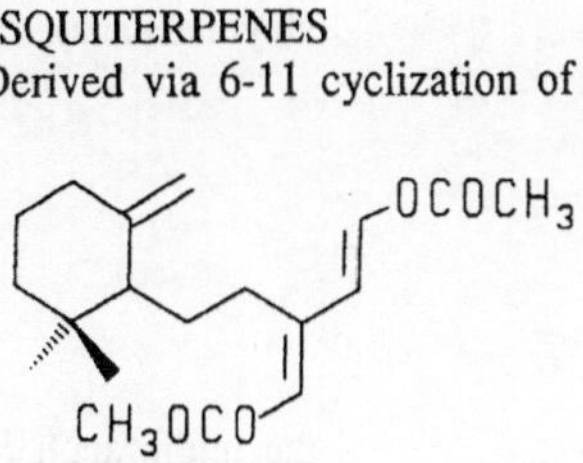

OCOCH₃

green seaweeds of the genus *Caulerpa* (Caulerpaceae); ichtyotoxic.

CH₃OCO

'caulerpatriene'

203

B5.3A1. Derived from monocyclofarnesanes

(+)-avarol

the sponge *Dysidea avara*
(Dictyoceratida); inhibitor
of the cell division of
fertilized sea urchin eggs
and antiviral (AIDS).

nakafuran-8

the sponge *Dysidea*
fragilis (Dictyoceratida)
of Hawaii; antifeedant.

B5.3B. Derived via 1-6 cyclization of a farnesyl precursor: cadinanes and pupukeananes

heritol

the roots of the
mangrove plant *Heritiera*
littoralis; ichtyotoxic
(other cadinanes are
found in tropical
alcyonaceans).

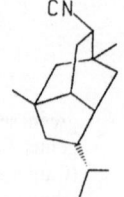

9-isocyanopupukeanane

the nudibranch *Phyllidia*
varicosa (Lamarck, 1801)
(Doridacea); antifeedant.

204

NATURAL PRODUCT STRUCTURES

B5.3C. Derived via 5-10 cyclization of a farnesyl precursor: penlanfuran

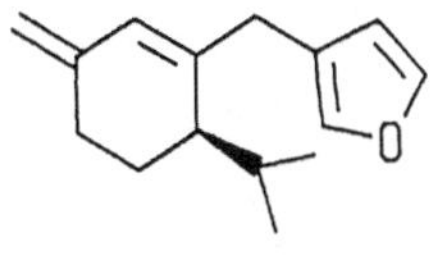

(-)-penlanfuran

the sponge *Dysidea fragilis* (Montagu, 1818) (Dictyoceratida) of Brittany.

B5.3D. Derived via cyclization of a germacrane precursor: eremophilanes and degraded eremophilanes

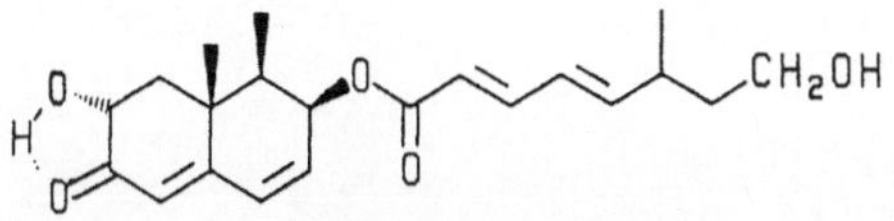

(+)-dendryphiellin-A

cultures of the deuteromycete *Dendryphiella salina*.

B5.3E. Other sesquiterpenes

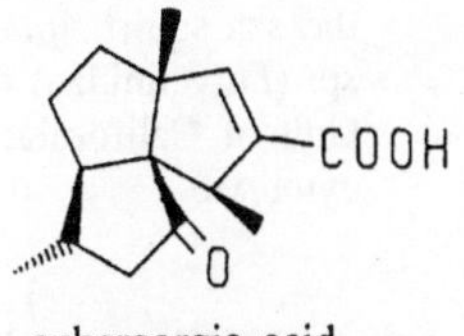

subergorgic acid

the gorgonian *Subergorgia suberosa*; cardiotoxic and inhibitor of neuromuscular transmission.

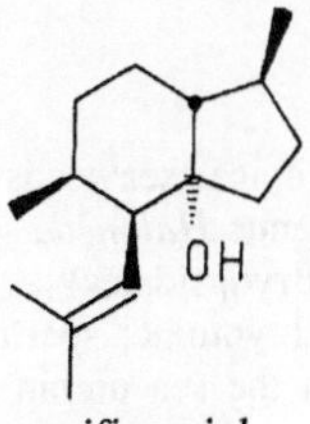

pacifigorgiol

the gorgonian *Pacifigorgia* cf. *adamsii*; ichtyotoxic (a similar terpene, tamariscol, occurs in the liverwort *Frullania tamarisci*).

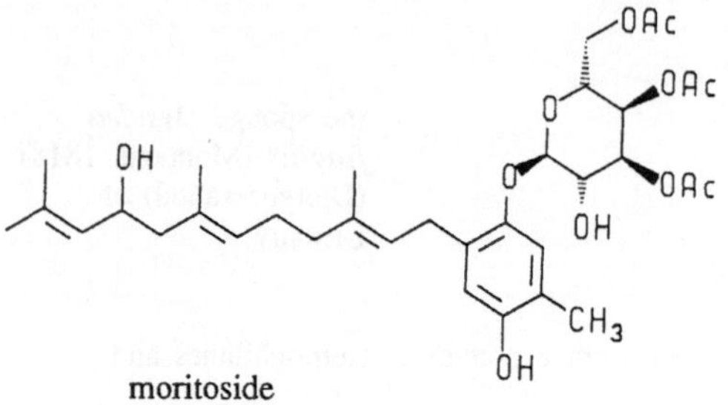

moritoside

the gorgonian *Euplexaura* sp. of Morito Beach, Gulf of Sagami; inhibitor of the first cell division of fertilized starfish eggs.

B5.4. DITERPENES
B5.4A. Linear diterpenes

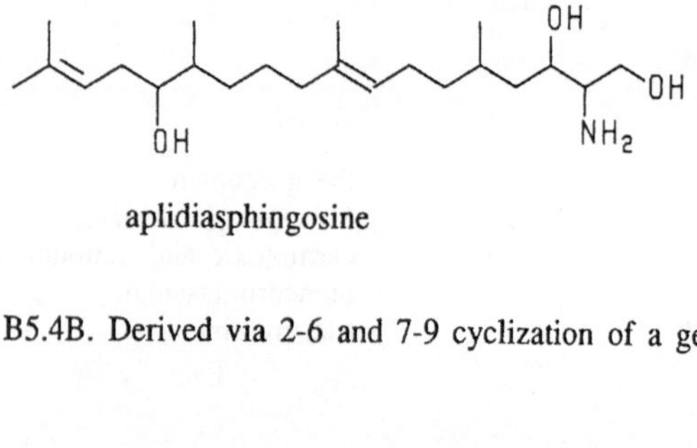

aplidiasphingosine

the sea squirt *Aplidium* sp. (Polyclinidae) of the Gulf of California; cytotoxic.

B5.4B. Derived via 2-6 and 7-9 cyclization of a geranyl geranyl (GG) precursor

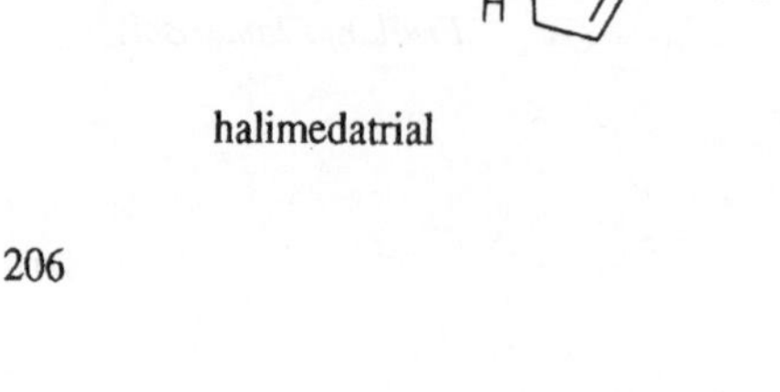

halimedatrial

tropical seaweeds of the genus *Halimeda* (Bryopsidales); ichtyotoxic; spermicidal in the sea urchin *Lytechinus pictus*; antibacterial and antifungal.

B5.4C. Derived via 2-10 cyclization of a GG precursor: degraded xenicanes

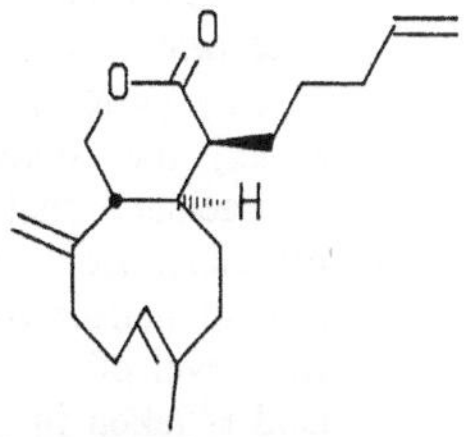

acalycixeniolide-A

the gorgonian
Acalycigorgia inermis;
inhibitor of the cell
division of fertilized eggs
of sea urchins; cytotoxic.

B5.4D. Derived via 1-14 cyclization of a geranyl geranyl precursor: the cembranes

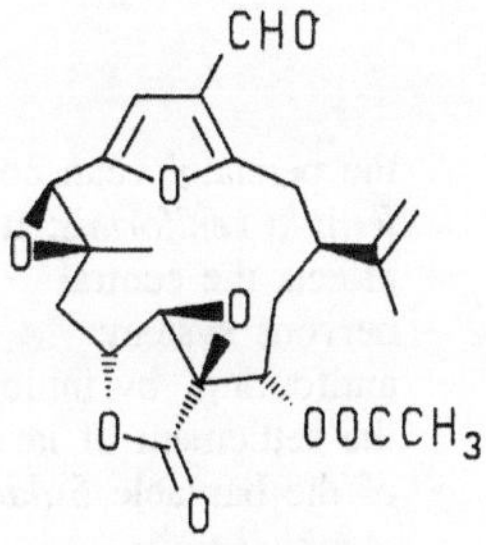

lophotoxin

the gorgonians
*Lophogorgia alba, L.
cuspidata, L. rigida*, and
L. chilensis; neuroactive,
used as a
neuropharmacological
tool.

B5.4E. Derived via 5-14 cyclization of a cembrane precursor: the cladiellanes

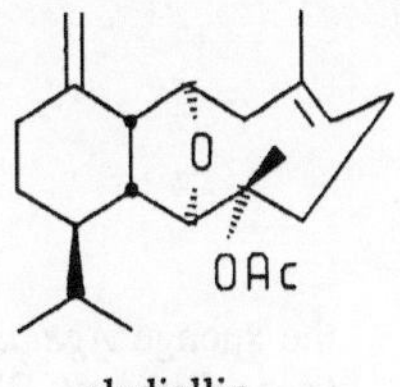

cladiellin

the alcyonacean *Cladiella*
sp. of the Great Barrier
Reef; antiinflammatory,
though toxic.

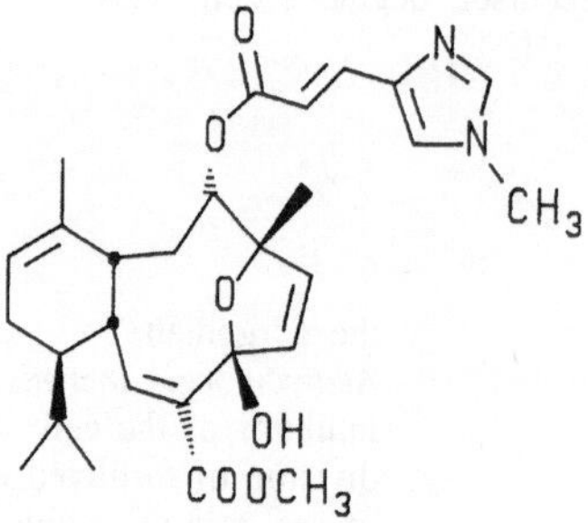

sarcodictyin-A

the stoloniferous octocoral *Sarcodictyon roseum* [= *Rolandia rosea*]; the estereal nitrogenated chain originates from urocanic acid of marine bacteria (the level of urocanic acid is taken in USA as a measure of spoilage of fish for market).

B5.4F. Derived via 8-13 cyclization of a cembrane precursor: the briaranes

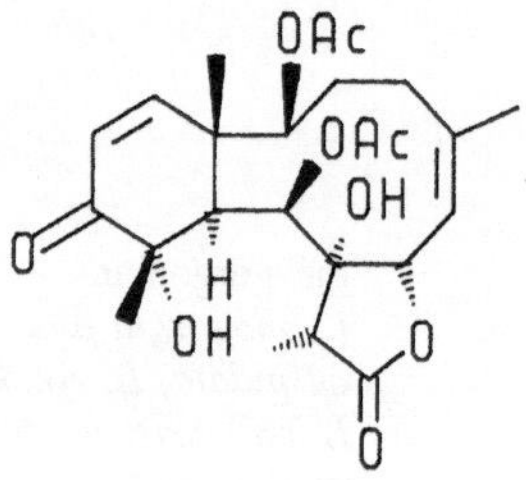

renillafoulin-A

the pennatulacean coral *Renilla reniformis*; it affects the central nervous system; antifouling by inhibiting the settlement of larvae of the barnacle *Balanus amphitrite*.

B5.4G. Prenylated 9,4-friedodrimanes

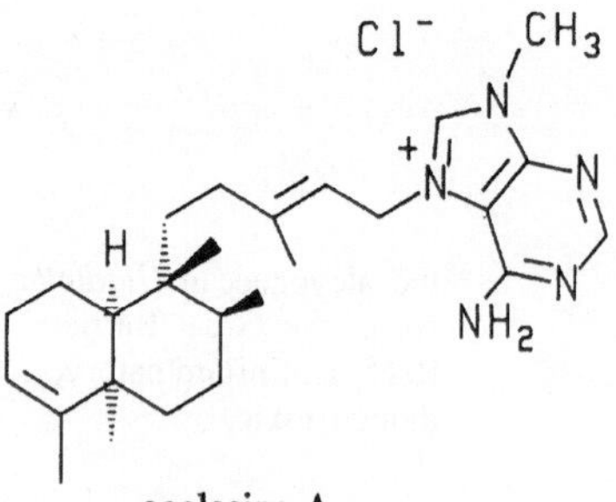

agelasine-A

the sponge *Agelas* sp. (Axinellida) of Okinawa; it inhibits Na⁺,K⁺-ATPase.

the sponge $Agelas$ sp. (Axinellida) of Okinawa; it inhibits Na^+,K^+-ATPase.

B5.4H. Other diterpenes

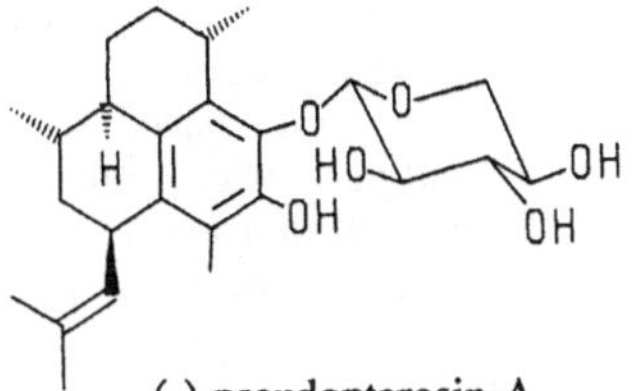

(-)-pseudopterosin-A

the gorgonian
*Pseudopterogorgia
elisabethae* (sp. nov.);
antiinflammatory.

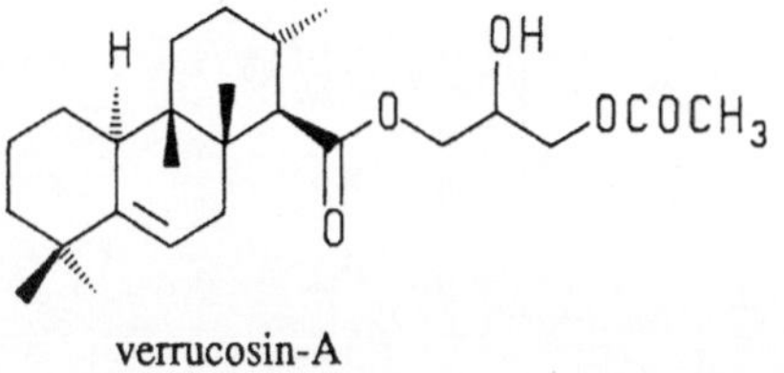

verrucosin-A

from the dorid
nudibranch *Doris
verrucosa*; ichtyotoxic.

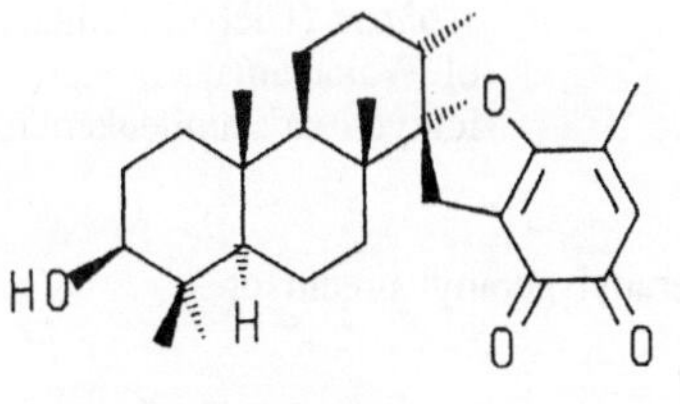

stypoldione

the brown seaweed
Stypodium zonale
(Lamoroux) (Dictyotales);
it inhibits the cell
division of fertilized sea
urchin eggs; cytotoxic;
ichtyotoxic.

B5.5. SESTERTERPENES
B5.5A Linear sesterterpenes

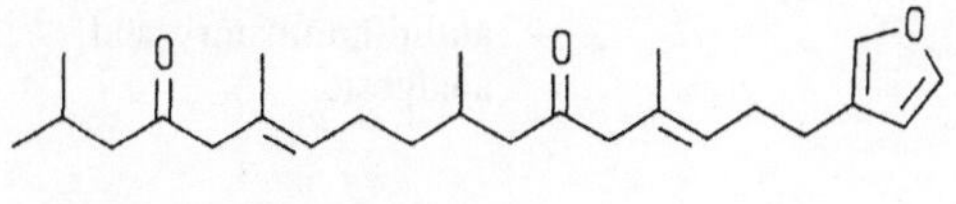

idiadione

the sponge *Spongia idia*
(de Laubenfels) [=
Leiosella idia]
(Dictyoceratida); toxic to
starfish, shrimps, and
bryozoans.

5.5B. Derived via Me(3)-7 cyclization of a geranyl farnesyl precursor

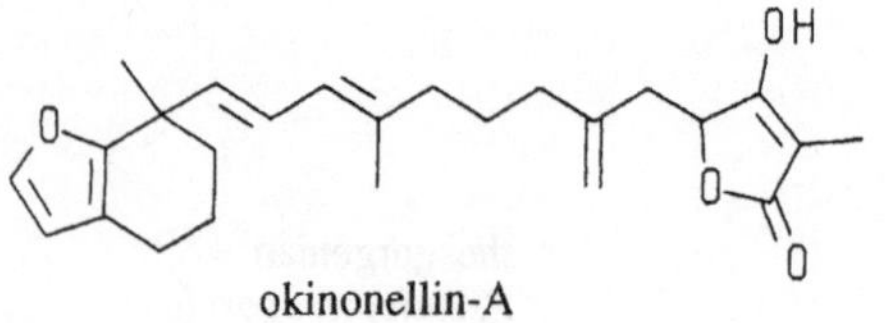

okinonellin-A

the sponge *Spongionella* sp. (Dictyoceratida) of Okinoshima Island; antispasmodic; it inhibits the cell division of fertilized sea urchin eggs.

B5.5C. Derived via 1-7, 6-11, 10-15, and 14-19 cyclization of a geranyl farnesyl precursor: the scalaranes

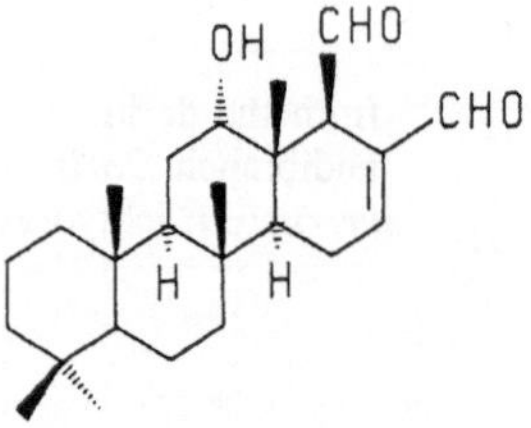

desacetylscalaradial

the sponge *Cacospongia scalaris* (Dictyoceratida) of Wakayama; ichtyotoxic; antileukemic.

B5.5D. Derived via prenylation of a monocyclogeranyl geranyl precursor

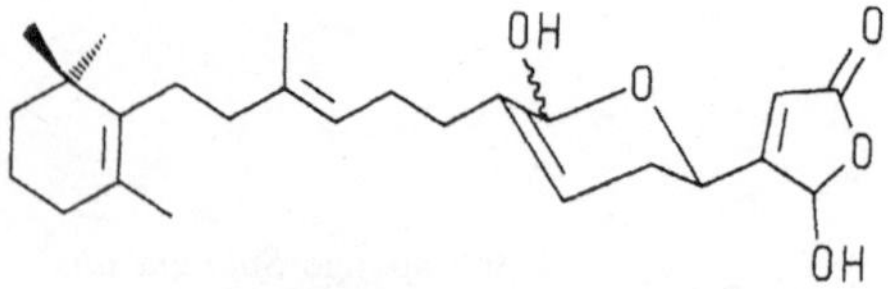

(4R)-manoalide

the sponge *Luffariella variabilis* (Polejaeff) (Dictyoceratida); antiinflammatory and analgesic.

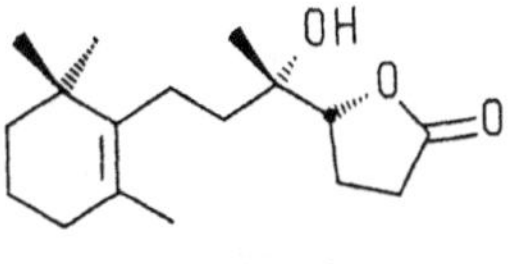

cavernosine

the sponge *Fasciospongia cavernosa* (Dictyoceratida); degraded terpene related to manoalide; ichtyotoxic.

B5.5E. Other sesterterpenes

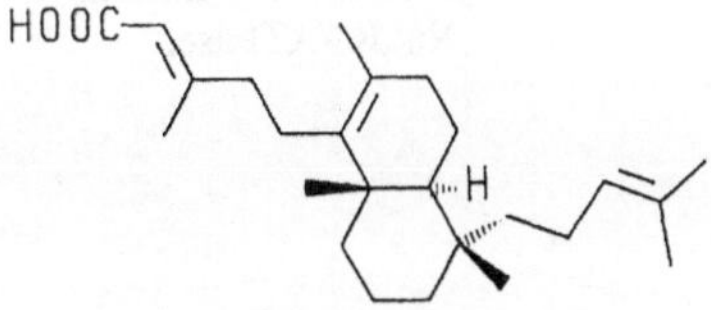

(+)-dysideapalaunic acid

the sponge *Dysidea* sp. (Dictyoceratida) of Palau; inhibitor of aldose reductase and therefore of interest in treating diabetes.

B5.6. TRITERPENES
B5.6A. Linear triterpenes

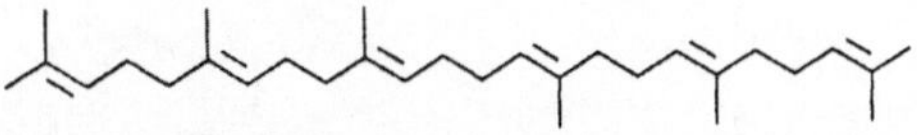

squalene

many marine animals, including shark (*Squalus*); precursor of steroids and triterpenes.

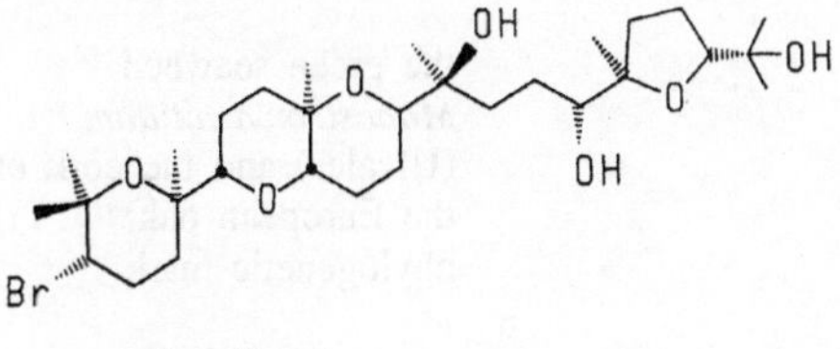

thyrsiferol

Laurencia thyrsifera (Hook) and *Laurencia obtusa* (red seaweeds, Ceramiales); cytotoxic.

211

B5.6B. Lanostane and cycloartane triterpenes

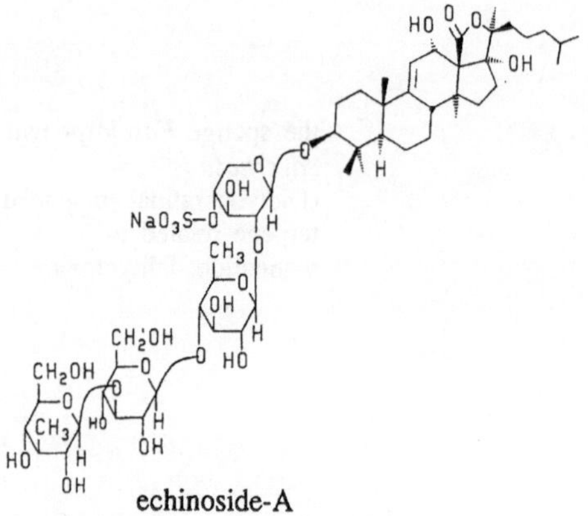

echinoside-A

the holothurian
Actinopyga echinites
(Jaeger); toxic deterrent;
hemolytic; alarm
substance; it inhibits
Na^+,K^+-ATPase.

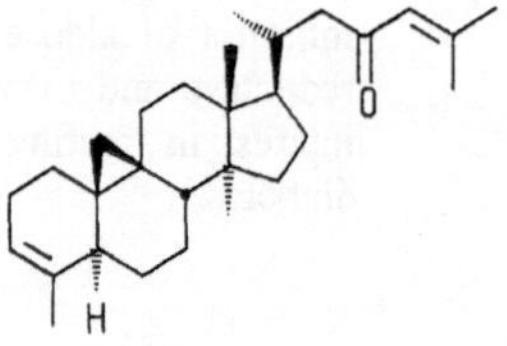

'tydemaniatriterpene'

the green seaweed
Tydemania expeditionitis
(Weber-van Bosse)
(Bryopsidiaceae);
phylogenetic mark.

B5.6C. Friedelane triterpenes

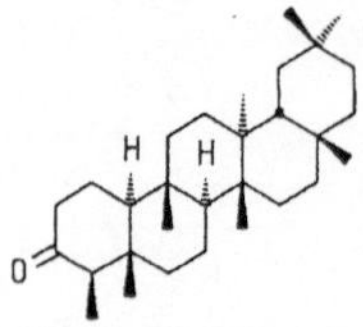

friedelin

the green seaweed
Monostroma nitidum
(Ulvales) and the cork of
the European oak;
phylogenetic mark.

NATURAL PRODUCT STRUCTURES

B5.6D Oleane triterpenes

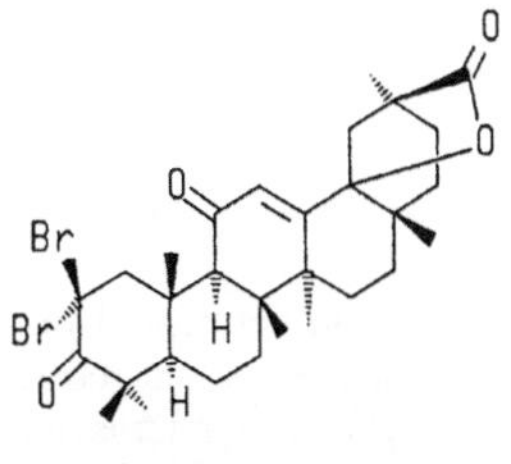

(20β)-echinolactone-B

the scleractinian coral *Echinopora lamellosa* (probably: Esper, 1797) (Faviidae); other oleanes occur in terrestrial flowering plants and petroleum of flowering-plant origin (sediments of Cretaceous or younger age).

B5.6E. Hopane triterpenes

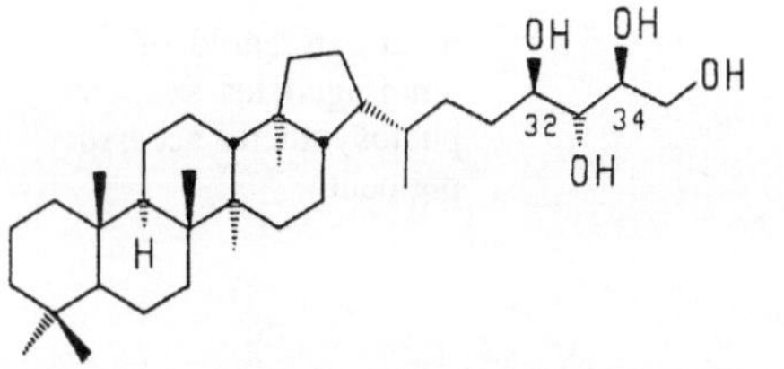

(32R,33R,34S)-bacteriohopanetetrol

it acts as a stabilizer of membranes of bacteria; also found in fossil sediments.

B5.6F. Other triterpenes

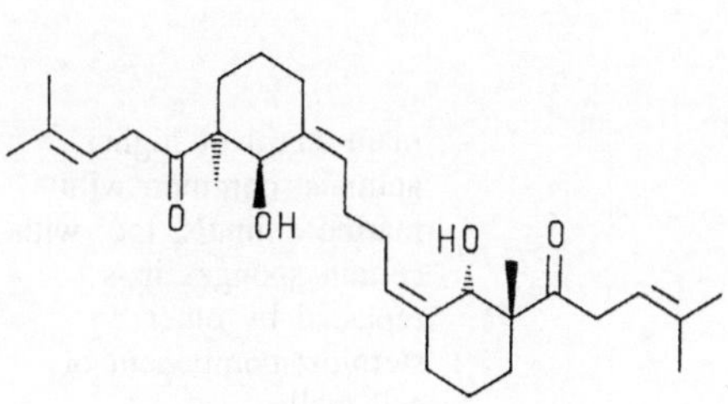

limatulone

the archeogastropod mollusc *Collisella limatula*; antifeedant.

B5.7. TETRATERPENES

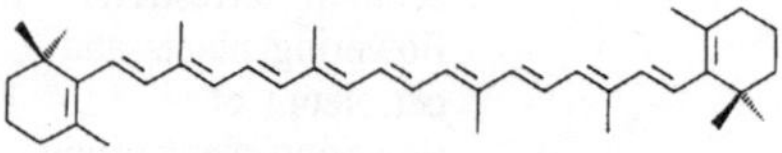

β,β-carotene

main carotenoid of seaweeds and terrestrial plants as photosynthetic accessory pigment.

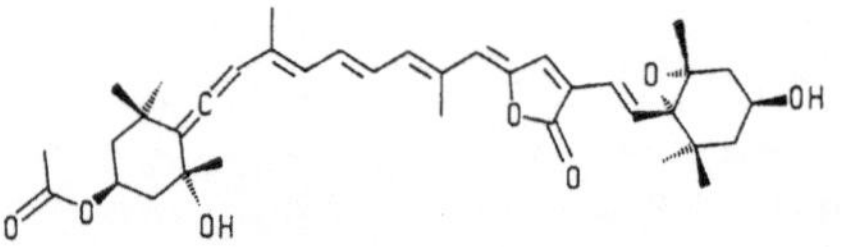

peridinin

main carotenoid of dinoflagellates as photosynthetic accessory pigment.

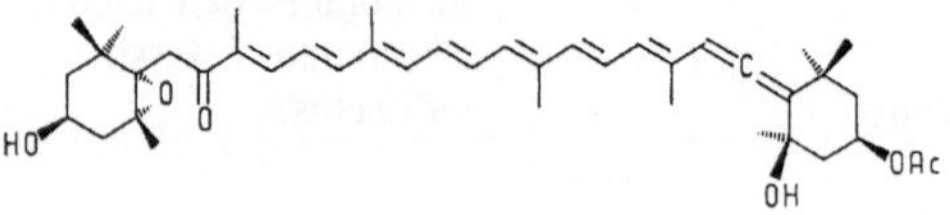

fucoxanthin

main carotenoid of diatoms as photosynthetic accessory pigment.

B6. STEROIDS
B6.1. Cholestanes

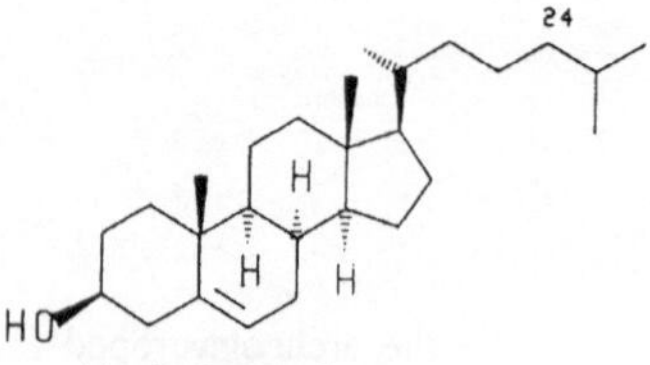

(-)-cholesterol

main sterol of higher animals; common with marine animals, too; with certain sponges it is replaced by other steroids; component of cell walls.

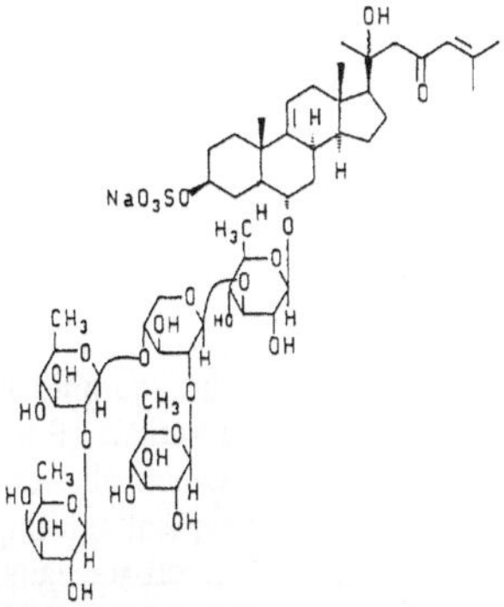

acanthaglycoside-A

the starfish *Acanthaster planci* (L.); toxic deterrent and alarm substance; hemolytic.

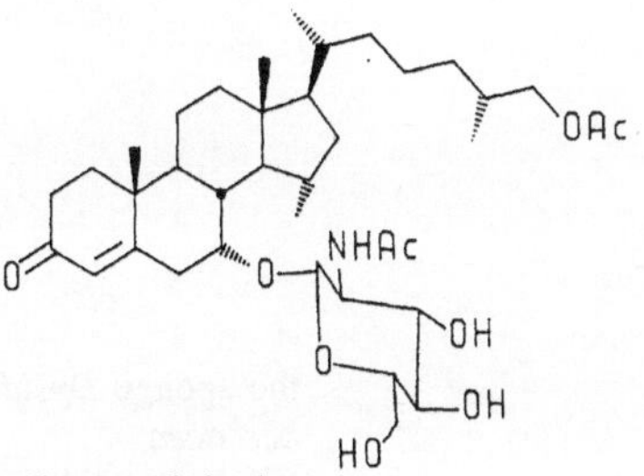

(-)-pavoninin-1

Pardachirus pavoninus (peacock sole); shark repelling; hemolytic.

B6.2. 4-Methylsteroids

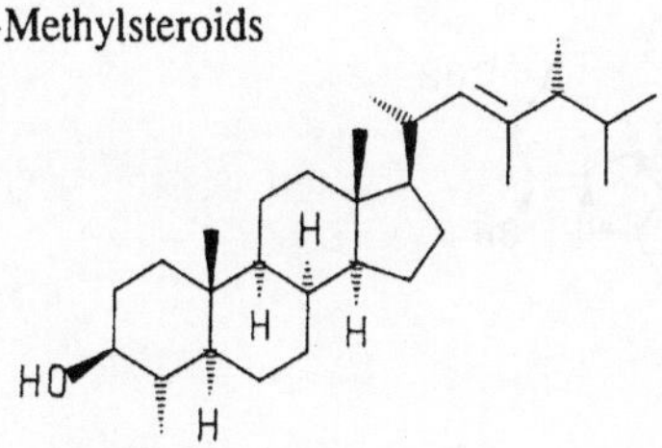

dinosterol

main sterol of the dinoflagellates; found also found in gorgonians; chemical mark.

215

B6.3. Ecdysteroids

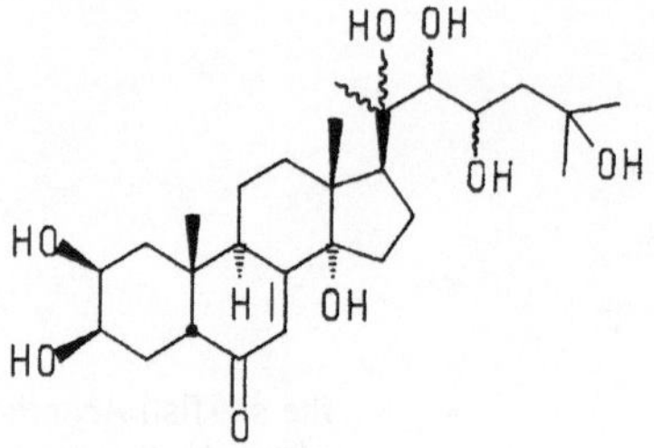

gerardiasterone

the zoanthid *Gerardia savaglia* (Bertolini, 1819) which also contains typical molting hormones of crustaceans and insects.

B6.4. Seco-steroids

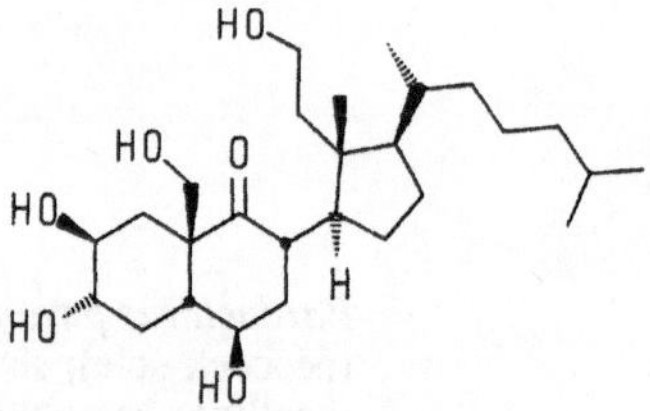

herbasterol

the sponge *Dysidea herbacea* (Dictyoceratida); ichtyotoxic.

B6.5. Dimeric steroids

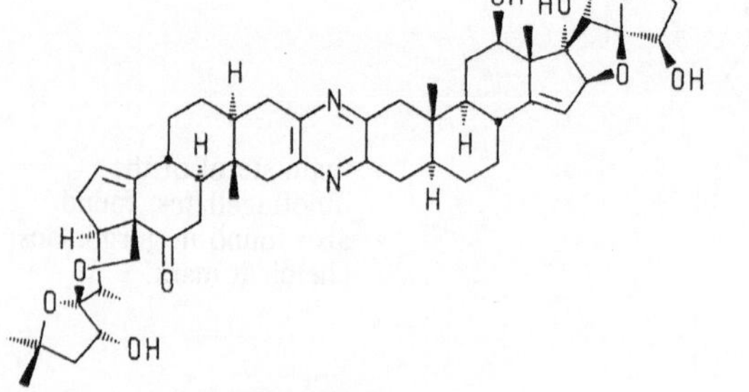

(+)-cephalostatin-I

the hemichordate *Cephalodiscus gilchristi*; antileukemic.

B7. NITROGEN COMPOUNDS
B7.1. AMINO ACIDS
B7.1A. Protein amino acids

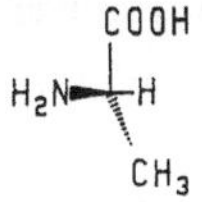

COOH
H$_2$N—H
    CH$_3$

L-alanine

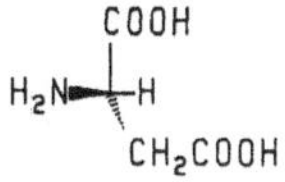

COOH
H$_2$N—H
    CH$_2$COOH

L-aspartic acid

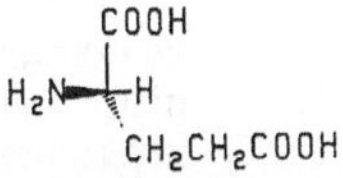

COOH
H$_2$N—H
    CH$_2$CH$_2$COOH

L-glutamic acid

main amino acids in
most seaweeds; protein
components.

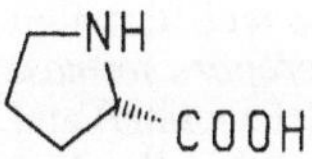

—NH
    ,,,,COOH

L-proline

main amino acid in
seaweeds of the family
Rhodomelaceae; chemical
mark.

B7.1B. Non-protein amino acids

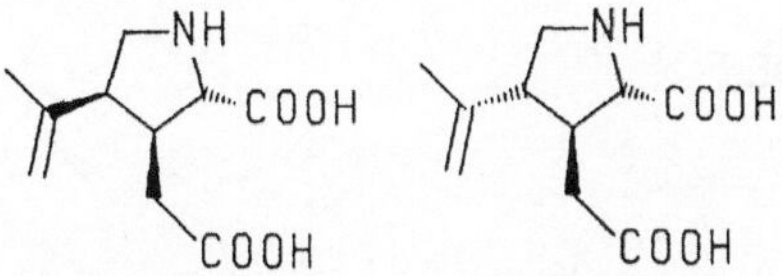

—NH
    ,,,,COOH
COOH

(-)-α-kainic acid

—NH
    ,,,,COOH
COOH

(+)-α-allokainic acid

Digenea simplex (red
seaweed, Ceramiales,
Rhodomelaceae);
insecticidal; anthelmintic;
neurobiological tools as
neuroexcitants.

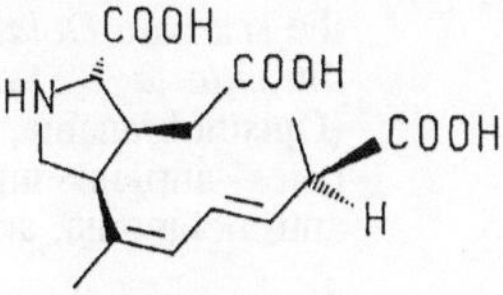

COOH
HN   COOH
            COOH
            ,,,H

(-)-domoic acid

Chondria armata (red
seaweed, Ceramiales,
Rhodomelaceae);
bioactivity similar to
α-kainic acid.

217

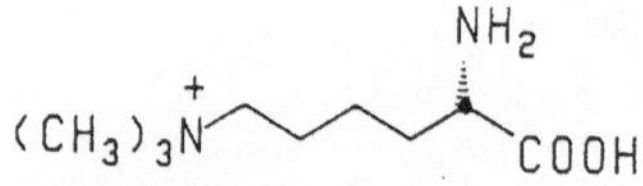

$(CH_3)_3\overset{+}{N}$ ⋯⋯ $\overset{NH_2}{\underset{}{\diagup}}$ COOH

laminine

H_2N ⋯⋯ COOH

GABA

homarine

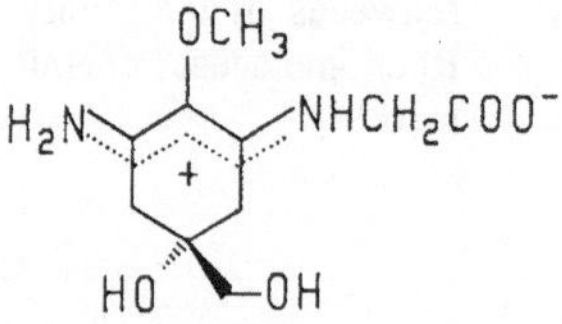

OCH$_3$

H_2N ⋯⋯ NHCH$_2$COO⁻

HO ⋯OH

palythine

Heterochordaria abietina [= *Analipus japonicus*] and other brown seaweeds of the genus *Laminaria* (Laminariales); vascular and muscular activities.

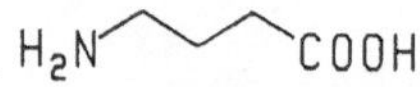

calcareous red seaweeds; it induces the settlement of mollusc larvae; neurotransmitter in man. many marine invertebrates, in particular the gorgonians *Leptogorgia virgulata* and *L. setacea*; antifouling.

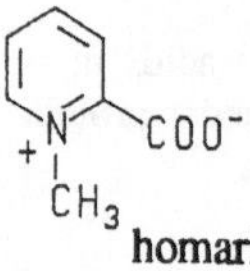

the reef-forming coral *Acropora formosa* (Scleractinia) and the zoanthid *Palythoa tuberculosa* as protective filter against solar radiations; similar compounds in the bivalve mollusc *Mytilus galloprovincialis*.

B7.2. PEPTIDES AND POLYPEPTIDES
B7.2A. Peptides

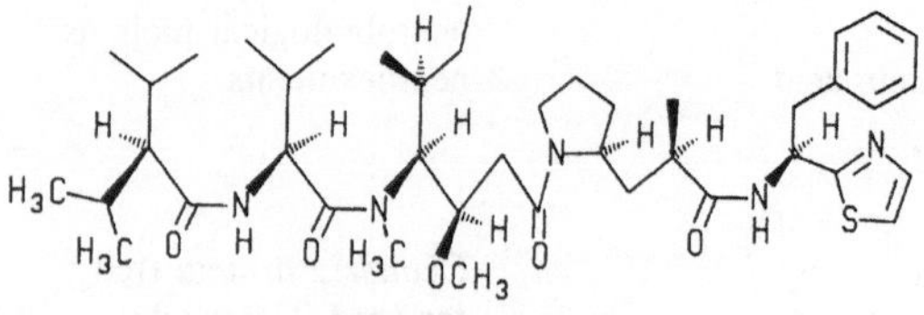

(-)-dolastatin-10

the sea hare *Dolabella auricularia* (Opisthobranchia, sea hares) antileukemic; anti-melanoma; antiviral.

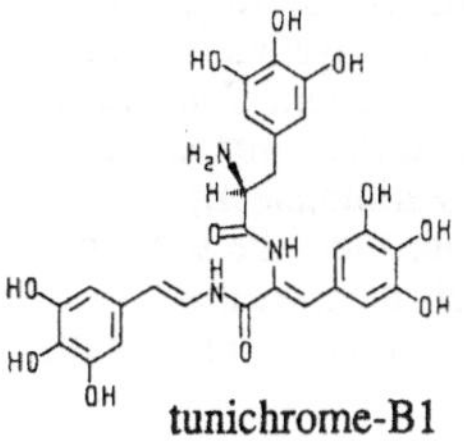

tunichrome-B1

the ascidian *Ascidia nigra* (L.) (Phlebobranchia); reducing and complexing agent for vanadium; possibly is involved in the immune system.

B7.2B Depsipeptides

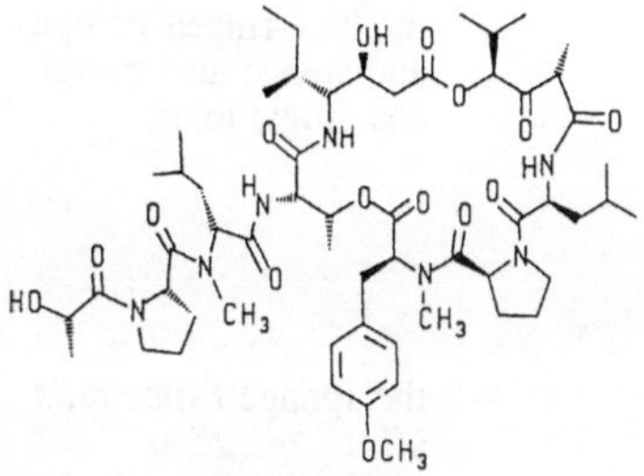

(-)-didemnin-B

the ascidian *Trididemnum solidum* (Van Name) (Aplousobranchia, Didemnidae); antitumoral and antiviral.

B7.3. ALKALOIDS
B7.3A. Alkaloids derived from arginine

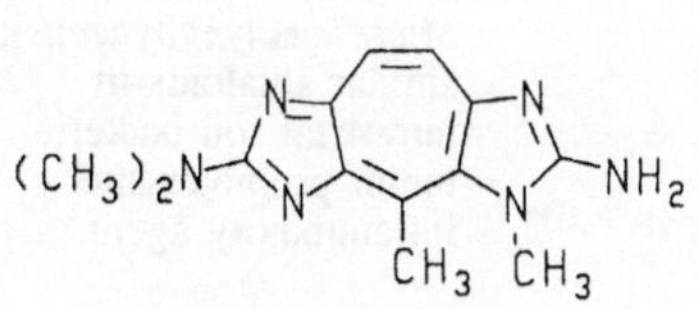

zoanthoxanthin

the zoanthids *Parazoanthus axinellae* and *P. axinellae adriaticus*; it intercalates into DNA.

219

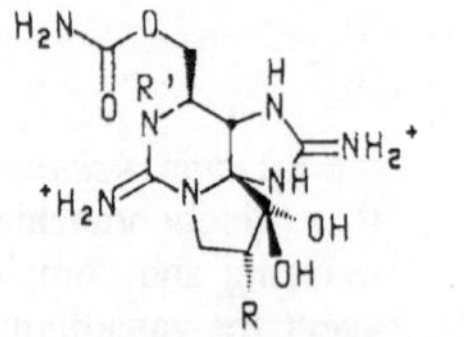

saxitoxin (R = R' = H)
gonyautoxin-1 (R = OSO₃⁻; R' = OH)

mollusks and fishes as dietary products of dinoflagellate origin (of the genera *Gonyaulax* and *Protogonyaulax* (Peridiniales)); it obstructs the sodium channels at the membranes of animals.

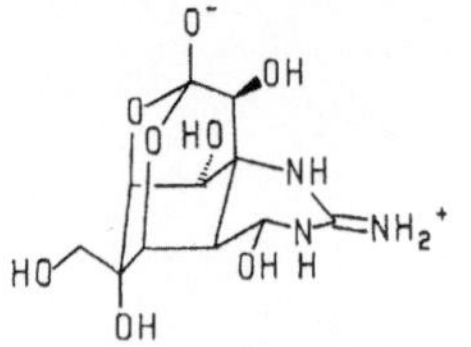

(-)-tetrodotoxin

marine bacteria, fish, starfish, xanthid crabs, the blue-ringed octopus, and snails; also newts and frogs; toxin.

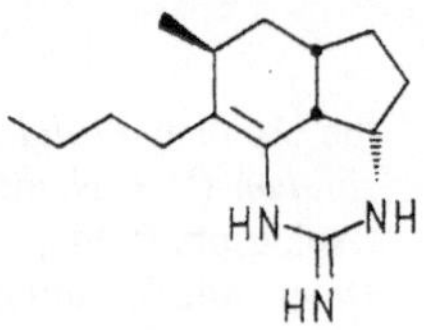

(+)-ptilocaulin

the sponge *Ptilocaulis* aff. *P. spiculifer* (Axinellida); antileukemic and antibacterial.

B7.3B. Alkaloids derived from tryptophan

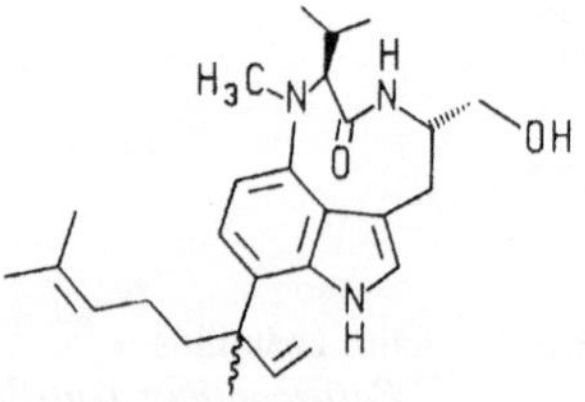

lyngbyatoxin-A

the cyanobacterium *Lyngbya majuscula* [= *Microleus lyngbyaceus*]; similar alkaloids in terrestrial soil bacteria; tumor promoter and inflammatory agent.

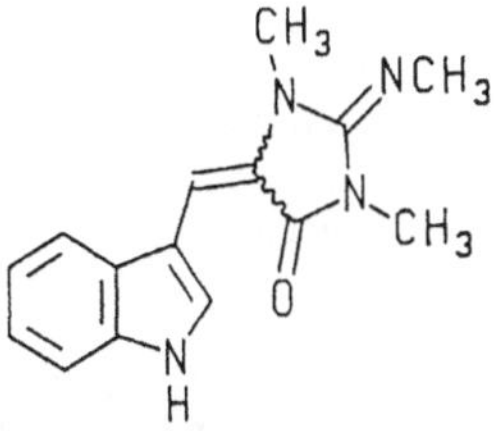

the sponges *Aplysinopsis reticulata* and *Smenospongia* [= *Polyfibrospongia*] *echina* (Dictyoceratida) and scleractinian corals (Dendrophylliidae); it affects neurotransmission; antidepressant; it alleviates noxious side effects of certain drugs; probably a protective filter against solar radiation.

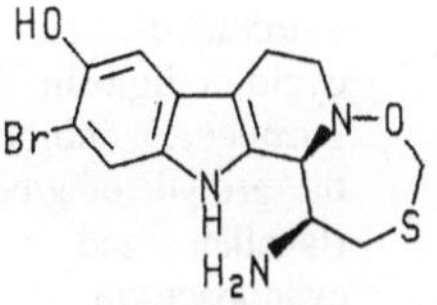

methylaplysinopsin

the ascidian *Eudistoma olivaceum* (Aplousobranchia); antiviral. Other β-carbolines also from the ascidian *Ritterella sigillinoides* (Brevin, 1956) and the hydroid *Aglaophenia pluma* (L.).

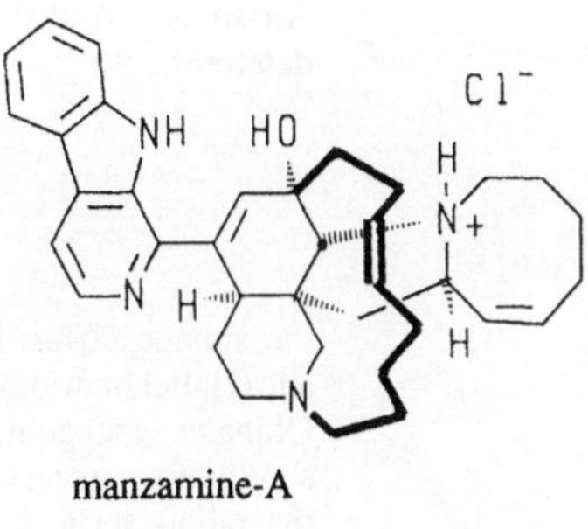

eudistomin-C
(O-α,β equilibrium forms)

manzamine-A

the sponge *Haliclona* sp. (Haplosclerida) collected off Manzamo Island, Okinawa; antileukemic.

221

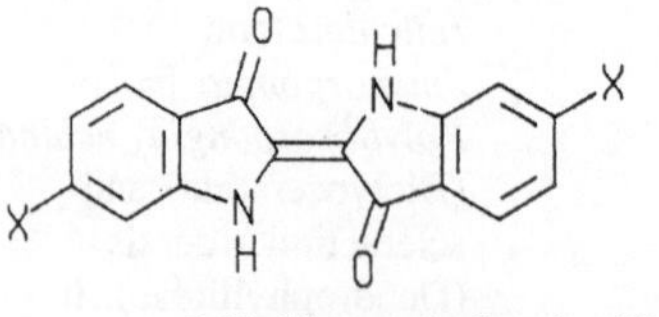

6,6'-dibromoindigotin (X = Br)
indigotin (X = H)

prosobranch mollusks of
the genera *Murex* and
Dicathais; dying pigment
stable to solar radiation.

B7.3C. Alkaloids derived from proline

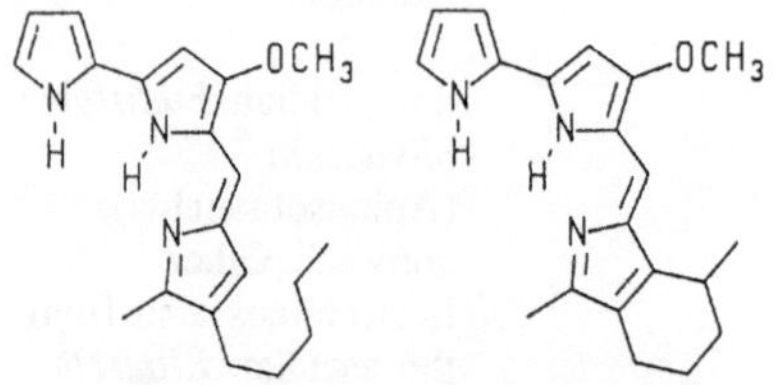

prodigiosin   cycloprodigiosin

marine bacteria
(prodigiosin: *Beneckea
gazogenes, Alteromonas
rubra*, and *Serratia
marcescens*;
cycloprodigiosin: *B.
gazogenes*); inhibitors of
the growth of green
flagellates and
cyanobacteria.

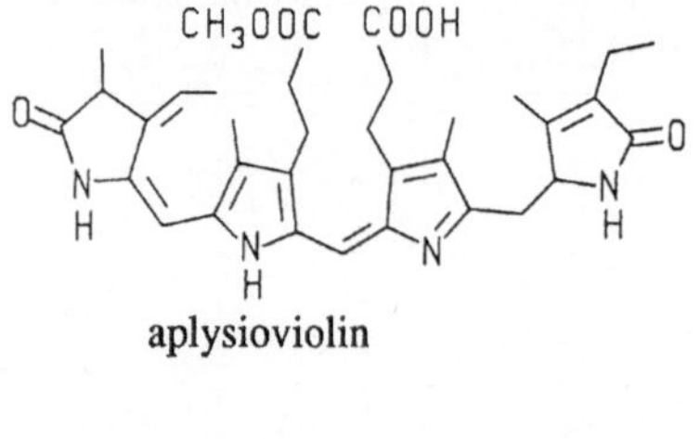

aplysioviolin

sea hares of the genus
Aplysia (Opisthobranchia,
Anaspidea, Aplysiidae);
deterrent.

B7.3D. Alkaloids derived from ornithine

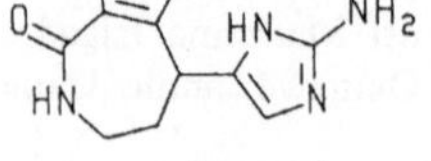

hymenin

the sponge *Hymeniacidon*
sp. (Halichondrida) of
Okinawa; antagonist of
serotonergic receptors in
the rabbit aorta; it blocks
α-adrenoceptor activity.

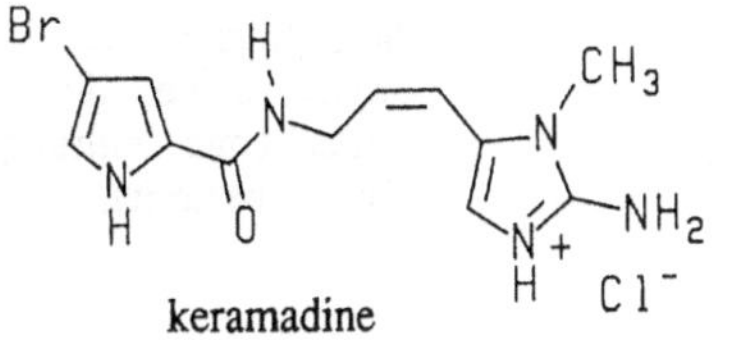

keramadine

the sponge *Agelas* sp. of
Kerama Retto
(Axinellida); antagonist
of serotonergic receptors.

B7.3E. Alkaloids derived from lysine

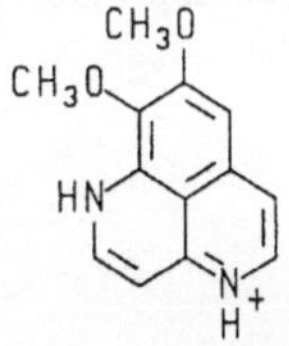

aaptamine

the sponge *Aaptos aaptos*
(Hadromerida); antagonist
of serotonergic receptors
in the rabbit aorta; it
blocks α-adrenoceptor
activity.

B7.3F. Alkaloids derived from phenylalanine

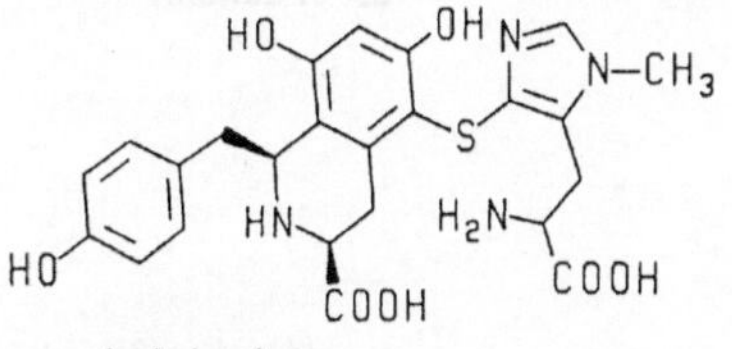

imbricatine

the starfish *Dermasterias
imbricata*; deterrent.

B7.3G. Imidazole alkaloids

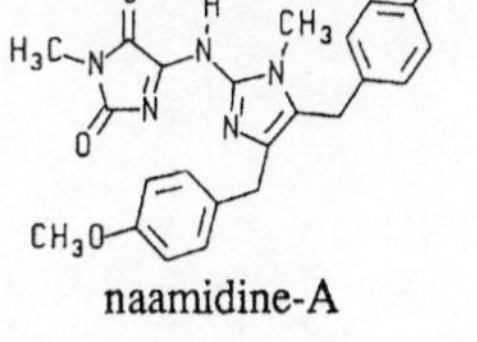

naamidine-A

the sponge *Leucetta
chagosensis* (Dendy)
(Calcarea); similar
alkaloids in the
Mediterranean calcareous
sponge *Clathrina
clathrus*.

223

B7.3H. Other alkaloids

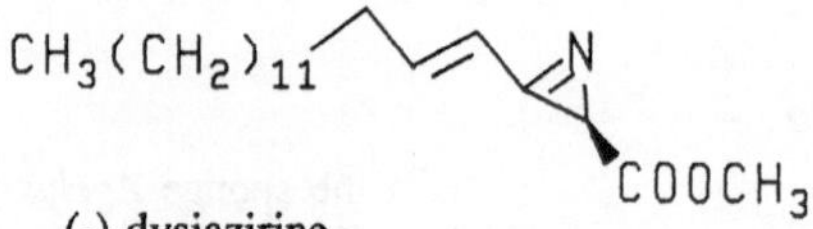

$CH_3(CH_2)_{11}$

(-)-dysiazirine

COOCH₃ → $COOCH_3$

the sponge *Dysidea fragilis* of the Fiji; antileukemic, antifungal, and antibacterial.

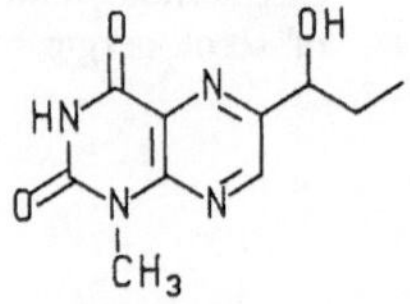

leucettidine

the sponge *Leucetta microraphis* (Calcarea); other pteridines are insect pigments.

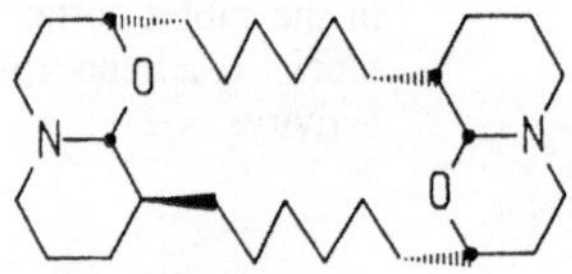

xestospongin-C

the sponge *Xestospongia exigua* (Petrosiida); vasodilatative.

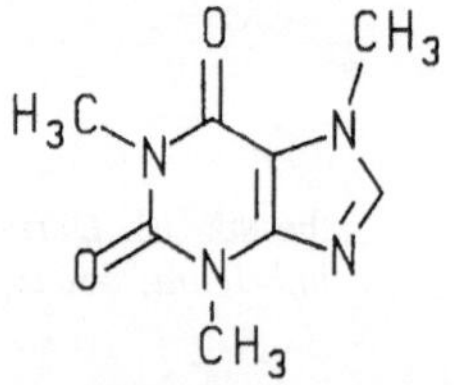

caffeine

in coffee beans and in the gorgonian *Paramuricea chamaeleon*; cardiotonic.

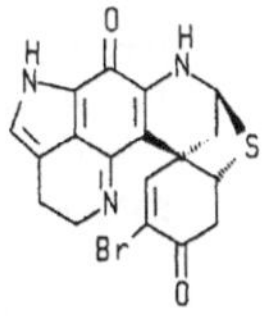

the sponges *Prianos melanos* (Halichondrida, Hymeniacidoniidae) and *Latrunculia brevis* (Ridley and Dendy) (Hadromerida); cytotoxic *in vitro* but not in vivo; toxic to mammals.

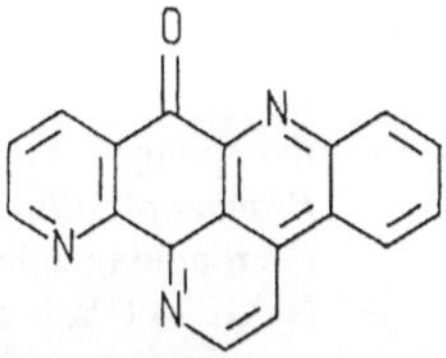

prianosin-A (= discorhabdin-A)

the ascidian *Didemnum* sp. (Aplousobranchia, Didemnidae); antileukemic; a Ca^{2+}-releaser in sarcoplasmic reticulum more potent than caffeine.

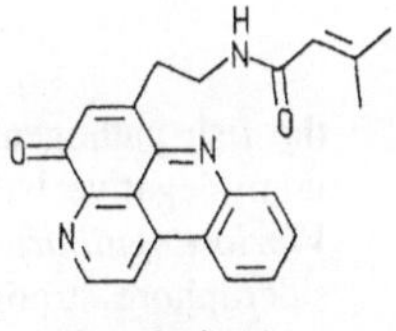

ascididemnin

the ascidian *Cystodytes dellechiajei* (Della Valle) (Aplousobranchia, Polycitoridae; bioactive as ascididemnin.

cystodytin-A

B7.4. OTHER AMINO ACID METABOLITES

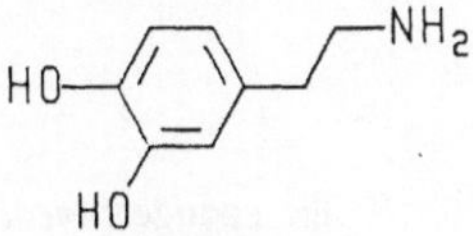

the alcyonaceans *Sinularia heterospiculata* and *Nephthea* sp.; atriastimulant; mediator in the central nervous system of man.

dopamine

225

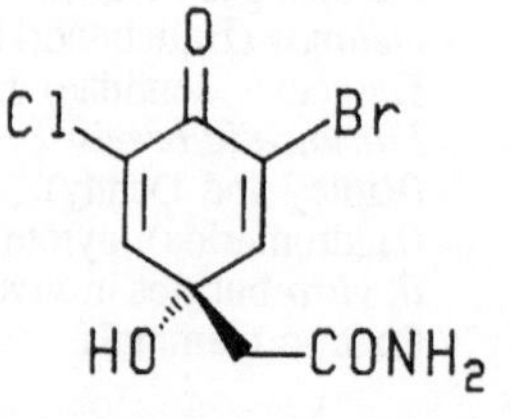

bromochloroverongiaquinol 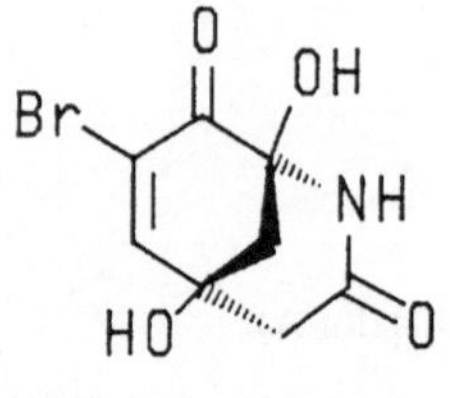   7-bromocavernicolenone

the sponge *Aplysina* [= *Verongia*] *cavernicola* (Vacelet) (Verongida); antibacterial; cytotoxic.

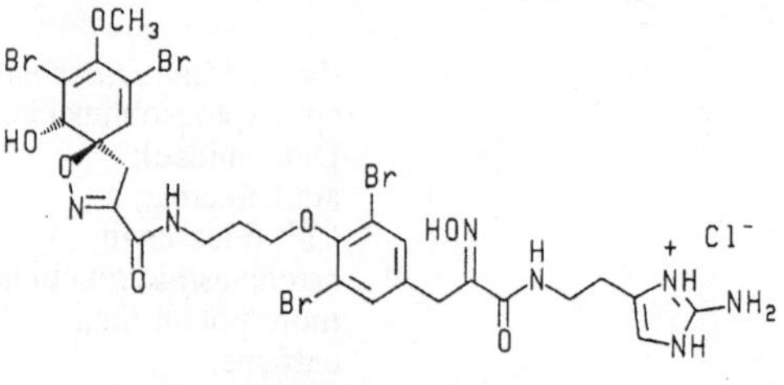

purealin

the sponge *Psammaplysilla purea* (Verongida); inhibitor of Na^+,K^+-ATPase and myosin Ca^{++}-ATPase; activator of myosin K^+,EDTA-ATPase.

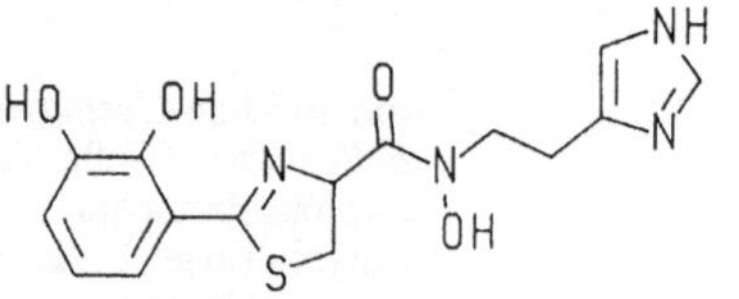

anguibactin

the fish pathogenic gram-negative bacterium *Vibrio anguillarum;* siderophore strongly involved in the virulence.

B7.5. NUCLEOSIDES

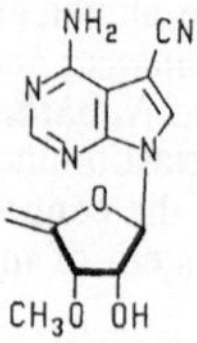

mycalisine-A

the sponge *Mycale* sp. (Poecilosclerida) of the Gulf of Sagami; inhibitor of the cell division of starfish eggs.

226

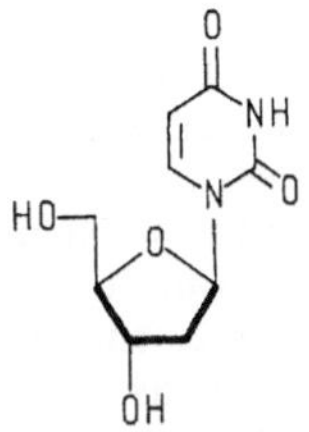

2'-deoxyuridine

the colonial ascidian
Trididemnum cereum
(Giard, 1872), starfish,
and acorn worms;
phylogenetic mark.

B8. SULFUR COMPOUNDS

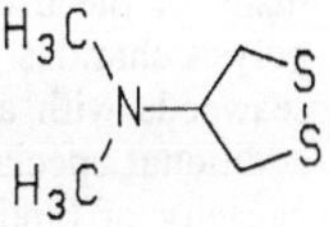

nereistoxin

the annelid *Lumbrinereis*
brevicirra [= *L.*
heteropoda]; insecticidal.

B9. BIOPOLYMERS
B9.1. POLYSACCHARIDES

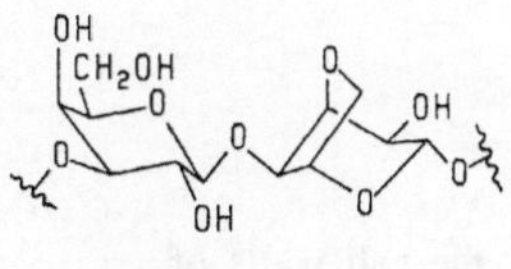

neutral agarose 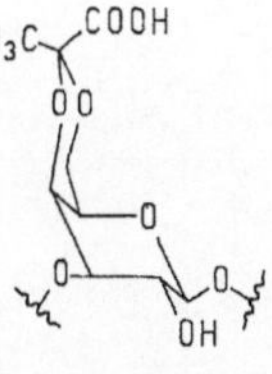   pyruvated agarose

red seaweeds; agarose is
used in immunological
studies; agar (made up of
agarose, pyruvated
agarose, and a sulfated
polymer of galactose) is
an emulsifying, binding,
suspending, and
thickening agent, used
also as a degradable
vehicle for oral drugs.

227

$$OSO_3^-$$
$$CH_2OH$$

K-carrageenan (R = H)
L-carrageenan (R = SO_3^-)

red seaweeds; gelling, emulsifying and stabilizing agents used for food and non-food, but especially for milky and waterish media; also used to stimulate the growth of connective tissues and to lower the cholesterol level in the blood.

D-mannuronic acid   L-guluronic acid

brown seaweed; alginic acids and alginates (made up of D-mannuronic and L-guluronic acid) have many of the uses of the polysaccharides of red seaweeds with an additional special capacity of binding water-soluble alkali-earth salts.

fucoidan (made of L-fucose and D-galactose in about 10:1 ratio)

brown seaweeds; base material for the production of L-fucose.

peptidoglycan
R = -CH(CH$_3$)C(O)-peptide. A single chain of two parallel chains is represented

the cell walls of gram-positive and gram-negative bacteria; structural polymers.

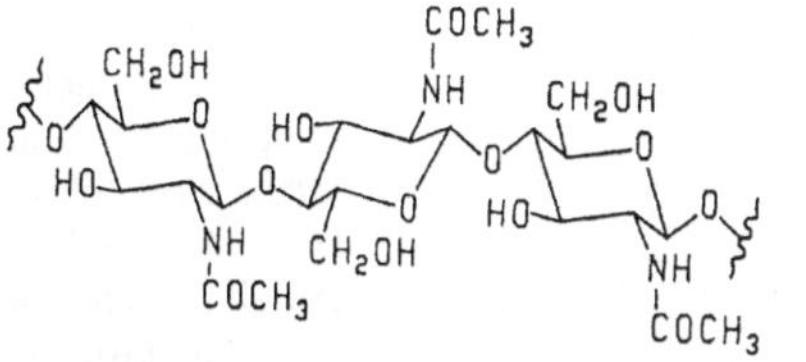

chitin

the cell walls of fungi
and, bound to a protein,
the cuticle of crustaceans;
structural polymer.

B9.2. OTHER BIOPOLYMERS

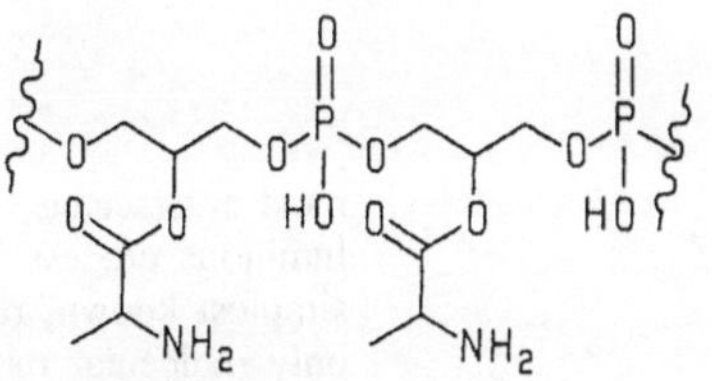

teichoic acids

the cell walls of
gram-positive bacteria;
structural polymers.

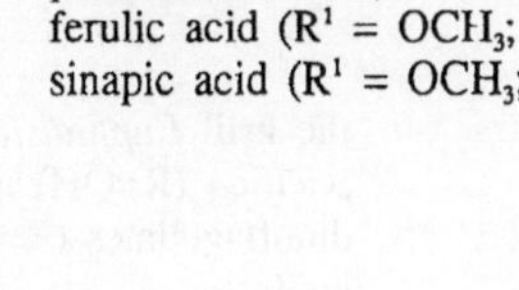

building blocks of lignins
p-coumaric acid ($R^1 = R^3 = H$; $R^2 = OH$)
ferulic acid ($R^1 = OCH_3$; $R^2 = OH$; $R^3 = H$)
sinapic acid ($R^1 = OCH_3$; $R^2 = R^3 = OH$)

seagrasses, emergent
plants, and terrestrial
plants; structural
polymers.

B10. LUCIFERINS

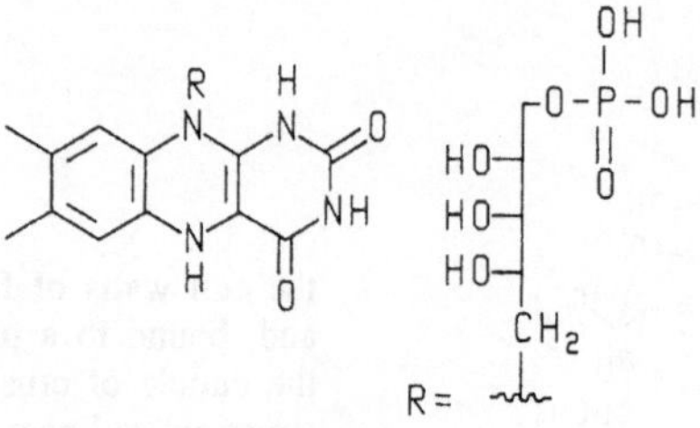

dihydroflavin monophosphate

marine bacteria; a long-chain aldehyde, molecular oxygen, vitamin-B_{12}, phosphoric acid, and the luciferase are also needed in the luminous process; luciferin.

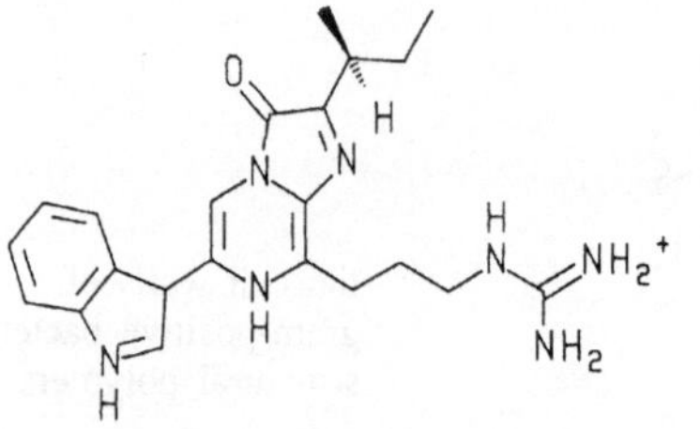

'cypridinaluciferin'

most crustaceans; the luminous process is the simplest known, requiring only molecular oxygen and the luciferase; luciferin.

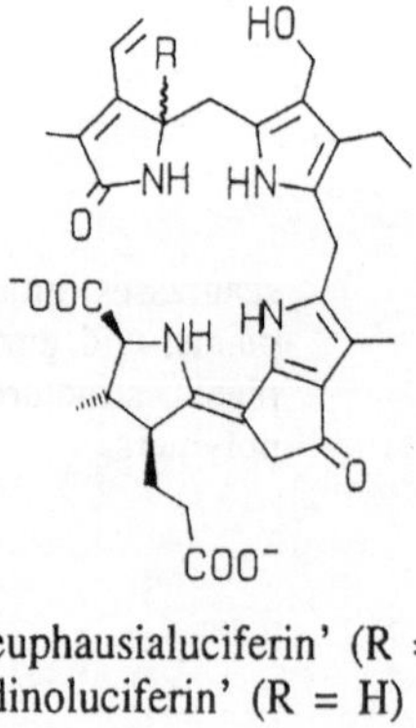

'euphausialuciferin' (R = OH)
'dinoluciferin' (R = H)

the krill *Euphausia pacifica* (R=OH) and dinoflagellates (R=H); luciferins.

NATURAL PRODUCT STRUCTURES

coelenterazine

Aequorea forskalea (hydroid); calcium and luciferase are also involved; unusually for cnidarians another protein is implicated in the luminous process while molecular oxygen is not; prosthetic group of this luciferin (aequorin).

C. Phylogenetic relationships among living organisms and natural products of phylogenetic significance

The biological classification of living organisms is generally controversial, and phylogenetic relationships is what specialists disagree on most. Until recently this was especially true for soft-bodied organisms, such as the ascidians, which have left no fossil records. Now opinions are changing: fossil records, which were held in great respect in establishing phylogenetic relationships, are no longer deemed informative enough to alter a phylogeny based on data for living species. This alleviates the problems with soft-bodied organisms while calling for revisions in phylogenetic assignments for the other organisms.

What I am presenting here is an attempt to establish phylogenetic relationships among living organisms on the basis of their natural products. Phylogenetic relationships based on proteins or genetic material are the best examples, and have been already dealt with in previous sections of this book, while secondary metabolites confront us with a most ambitious task. This is apparent from the frequent isolation of the same secondary metabolites from taxonomically unrelated species. In some cases this phenomenon is clearly the result of retention from the diet or production by symbionts and therefore there is no danger of deducing false phylogenetic relationships. Examples of dietary metabolites from previous chapters in this book concern antifeedant or toxic compounds found in opisthobranch mollusks, and toxins found in filter feeding mollusks. Examples of natural products which have symbiontic origin are okadaic acid (which is produced by the dinoflagellate *Prorocentrum lima* and accumulated in the sponges *Halichondria okadai* and *Halichondria melanodocia*) and dinosterol (which is produced by the dinoflagellate *Zooxanthella microadriatica* and transferred to gorgonians).

In other cases a dietary or symbiontic origin of the natural products are likely but not proven. This is exemplified by acanthifolicin (which is found in the sponge *Pandaros acanthifolium* and has a molecular structure very similar to that of okadaic acid), (20ß)-echinolactone-B (found in the scleractinian coral *Echinopora lamellosa*), domoic acid (found in both the red seaweed *Chondria armata* and mussels cultured in Cardigan Bay, Prince Edward Island, Canada), latrunculin-A (present in the sponge *Latrunculia magnifica* of the Red Sea and the nudibranch *Chromodoris elisabethina* of Guam and Eniwetok), and methylaplysinopsin (first found in demosponges and lately in scleractinian corals).

In certain cases neither a dietary nor a symbiontic origin can be advocated to rationalize the presence of the same secondary metabolite in taxonomically unrelated organisms; e.g. prostaglandin (15S)-PGA$_2$ (which is found as such in the semen of mammals and as methyl ester in the gorgonian *Plexaura homomalla* forma *kükentali*), ecdysterone (which is found in large amounts in the zoanthid *Gerardia savaglia* and as a hormone in crustaceans and insects), ajugasterone-C (a phytoecdysteroid first found in the terrestrial oriental plant *Ajuga* sp. and lately in the Mediterranean zoanthid *Gerardia savaglia*), caffeine (long known to be present in coffee seeds and recently found in the gorgonian *Paramuricea chamaelon*), and GABA (a primary metabolite in man and a settlement inducer in calcareous red seaweeds).

The danger of building up false phylogenetic conclusions through natural products, which is implicit in some of the above examples, can be minimized through a proper analysis. Let me first of all remark that primitive animals, such as the sponges, are not limited to dietary products but are also capable of *de novo* synthesis of secondary metabolites[12]. Such metabolites may safely be used in establishing phylogenetic relationships. In the subtle cases of sponge products of symbiontic origin there is a growing awareness that when obligate symbiosis is involved, as is frequently the case, the macroorganism-symbiont couple behaves, in what regards the synthesis of natural products, as if it were a single organism, irrespective of location and season, or a reproducible trend with changing seasons is

---

[12] *De novo* synthesis of conventional and unusual sterols occurs, for example, in the marine sponges *Microciona prolifera* and *Tethya aurantia californiana*.

233

observed. This applies to sponge sterols which have the C(24)-alkyl(alkenyl) chain typical of plant sterols and are in fact synthesized by the sponge algal symbionts. Such metabolites may therefore be used for taxonomic purposes.

In the following chart organisms which are regarded as equal in rank appear on the same horizontal line even if they belong to different phyla. Phylogenetic relationships are indicated by either an upward arrow or by a horizontal double arrow. Phylogenetically significant natural products appear enclosed in a box either at the right side or below the group of organisms of pertinence.

The chart (which does not take into account rRNA sequences because they are too short so far) illustrates that everything began with the primeval broth (molecular hydrogen, water vapor, carbon dioxide, nitrogen, ammonia, and methane) which formed the early atmosphere of the Earth. Under the action of the sun and of lightning, the primeval broth formed the amino acids which dissolved into the oceans and formed the proteins.

Although it is not yet clear how, supramolecular organization of the proteins and of genetic material gave origin to the cells of the simplest bacteria from which life evolved into either the autotrophic bacteria[13], capable of non-oxygenic photosynthesis, or the more complex cyanobacteria which are capable of oxygenic photosynthesis; the latter made molecular oxygen available for further evolution of living organisms. The cyanobacteria gave origin to the red algae along an endosymbiont line.

The green algae originated from photosynthetic bacteria via the prokaryotic green alga *Prochloron didemni* and gave origin to the higher plants which inherited the ability of their phylogenetic precursors to synthesize cycloartane or, as in the case of the European oak, friedelane triterpenes.

The ancestral animals, the protozoans, originated from the algae, too. In fact certain groups of algae, like the dinoflagellates, that have the typical feeding modes

---

[13]It has been long debated whether life originated from hydrothermal vents at the oceanic ridge (which are presented in chapter 4). Now this possibility has been dismissed; the temperatures there are too high.

PHYLOGENETIC RELATIONSHIPS

of animals, are traditionally in the domain of both the zoologist and the botanist.
The protozoans gave origin to the sponges, and probably also to the cnidarians

Origin of, and phylogenetic relationships among living organisms,
and natural products of phylogenetic significance

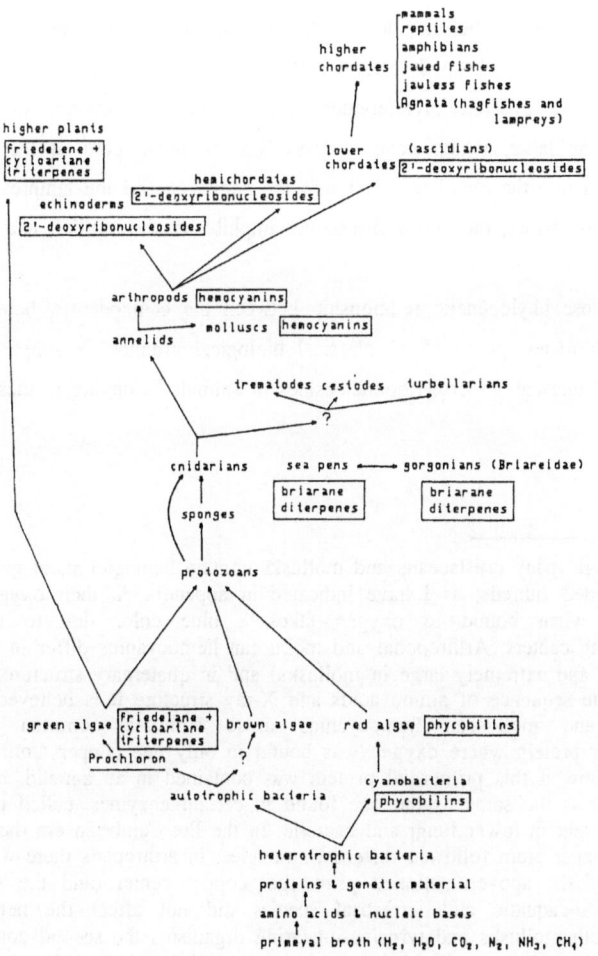

which may have descended from the sponges as well. Within the cnidarians, gorgonians of the family Briareidae and sea pens are interrelated by the production of briarane diterpenes.

Worms originated from cnidarians following two different lines of evolution. One line led, via an undefined intermediate organism, to the trematodes, cestodes, and turbellarians. The other line led to the annelids from which the large phyla of the mollusks and the arthropods originated[14]. The arthropods are the branching point before the more evolved invertebrates (echinoderms, hemichordates, and lower chordates). The latter, which have a dorsal hollow nerve cord at the embryonal stage, evolved into the chordates, first into the hagfishes and the lampreys and then into the jawless fishes, the jawed fishes, the amphibians, the reptiles, and finally the mammals.

The close phylogenetic relationship between the echinoderms, hemichordates, and lower chordates, proposed on classical biological grounds, is supported by the production of unusual 2'-deoxyribonucleosides in animals belonging to these groups.

---

[14]Although many crustaceans and mollusks contain hemoglobin, in general these are blue-blooded animals; as I have indicated in appendix A, their oxygen carrier, hemocyanin, when bound to oxygen takes a blue color due to the Cu(II) oxygen-binding centers. Arthropodal and molluscan hemocyanins differ in size (large in arthropods and extremely large in mollusks) and in quaternary structure. From the analysis of the sequence of amino acids and X-ray structure it is believed that both arthropodal and molluscan hemocyanins arose from a common primordial oxygen-carrier protein where oxygen was bound to only one copper atom. Although we do not know if this primordial protein was contained in an annelid, the binding site of copper is the same as the one found in certain enzymes, called tyrosinases, which are present in lower fungi and bacteria. In the Pre-Cambrian era the uptake of the second copper atom followed different strategies: in arthropods there was perhaps a doubling of the above-mentioned primordial copper center (and the scission of arthropods into aquatic and terrestrial species did not affect the hemocyanins) whereas in both mollusks and tyrosinase-bearing organisms the second copper center was built *ex novo*. Why mollusks have developed, and saved, such large proteins as oxygen carriers is a mystery.

REFERENCES AND SUGGESTIONS FOR FURTHER READING

biological aspects

Algae and plants

den Hartog, C. "The seagrasses of the world" North-Holland, Amsterdam, 1970.

Taylor, F.J.R, Ed. "The biology of dinoflagellates" Blackwell Scientific, London, 1987.

van den Ende, H. "Sexual interactions in plants" Academic Press, London, 1976.

Bacteria and fungi

Clayton, R.K.; Sistron, W.R., Ed. "The photosynthetic bacteria" Plenum Press, New York, N. Y., 1978.

Johnson, T.W.; Sparrow, F.K. "Fungi in oceans and estuaries" Cramer, 1961; reprinted by Wheldon & Wesley, Codicote, Herts, 1970.

Kohlmeyer, J.; Kohlmeyer, E. "Marine mycology" Academic Press, New York, N.Y., 1979.

Sieburth, J.McN. "Sea microbes" Oxford University Press, New York, N.Y., 1979.

Cnidarians

Joubin, L. "Le fond de la mer" Librairie Hachette, Paris, 1920.

Lacaze-Duthiers, H. "Histoire de la *Laura gerardiae*, type nouveau de crustacè parasite" Firmin-Didot, Paris, 1882.

Evolution

Schubert, I. "Eukaryotic nuclei of endosymbiontic origin" *Naturwissenshaften*, 75, 89, 1988.

REFERENCES

Echinoderms

Moran, P.J. in "Oceanography and marine biology" Aberdeen University Press, vol. 24, 1986.

General treatises

George, J.D.; George, J.J. "Marine life. An illustrated encyclopedia of invertebrates in the sea" Lionel Loventhal Ltd., London, 1979; published in USA by J. Wiley & Sons, New York, N.Y. Taxonomy and general biological aspects of marine invertebrates.

Parker, S.P. ed., "Synopsis and classification of living organisms" McGraw-Hill, New York, N.Y., 1982. Encyclopedic treatise on the taxonomy and general biological aspects of all living organisms.

History of coral reefs and theories about life

Daly, R.A. "The changing world of the ice age" Yale University Press, New Haven, 1934.

Darwin, C.R. "The structure and distribution of coral reefs", Smith, Elder & Co., London, 1842.

Hoffmeister, J.E.; Ladd, H.S. "The antecedent-platform theory" *J. Geol.*, 52, 388, 1944.

Lovelock, L. "Gaia: a new look at life on Earth" Oxford University Press, Oxford, 1979.

Wiens, H.J. "Atoll environment and ecology" Yale University Press, New Haven and London, 1971.

REFERENCES

Pharmacology

Baslow, M.H. "Marine pharmacology" Krieger, Huntington, New York, N.Y., 1977.

Hoppe, H.A.; Levring, T.; Tanaka, Y., eds., "Marine algae in pharmaceutical sciences" de Gruyter, Berlin, 1979.

Lewis, D.A. "Anti-inflammatory drugs from plants and marine sources" Birkhäuser, Basel, 1989.

Ragelis, E.P. "Seafood toxins" ACS Symposium Series 262, American Chemical Society, Washington D.C., 1984.

Wong, K. C.; Wu, L.T. "History of Chinese medicine" 2nd ed., National Quarantine Service, Shangai, 1936.

Scientific cruises

Challenger, H.M.S. "Reports of the scientific results of the exploring voyage H.M.S. Challenger", 1872-76.

Sponges

Bergquist, P.R. "Sponges" University of California Press, Berkeley and Los Angeles, 1978.

Levi, C. "Systématique de la classe des Demospongiaria" in Grassé P.P., ed., "Traité de Zoologie" Tome III, Fascicule I, pp. 577-631, Masson et C^{le}., Paris, 1973.

Simpson, Y.L. "The cell biology of sponges" Springer-Verlag, New York, N.Y., 1984.

Tuzet, O. "E'ponges calcaires" in Grassé P.P., ed., "Traité de Zoologie" Tome III, Fascicule I, pp. 27-132, Masson et C^{le}., Paris, 1973.

Tuzet, O. "Hexactinellides ou hyalosponges" in Grassé P.P., ed., "Traité de Zoologie" Tome III, Fascicule I, pp. 633-690, Masson et C^{le}., Paris, 1973B.

REFERENCES

natural product chemistry

Christophersen, C. "Marine alkaloids" in "The alkaloids" Vol. XXIV, Academic Press, New York, N.Y., 1985

Dedekind, A. "Ein Beitrag zur Purpurkunde. Nebst Anhang: Neue Ausgaben seltener älterer Schriften über Purpur" Meyer und Müller, Berlin, 1898.

Faulkner, D.J. "Marine natural products: metabolites of marine algae and herbivorous marine mollusks" *Nat. Prod. Rep.*, *1*, 251, 1984; "Marine natural products: metabolites of marine invertebrates" *ibid.*, *1*, 551, 1984; "Marine natural products" *ibid.*, *3*, 1, 1986; *ibid.*, *4*, 539, 1987; *ibid.*, *5*, 613, 1988.

Garson, M.J. "Biosynthetic studies on marine natural products" *Nat. Prod. Rep.*, *6*, 143, 1989.

Hashimoyo, Y. "Marine toxins and other bioactive marine metabolites" Japan Scientific Societies Press, Tokyo, 1979.

Hegnauer, R. "Chemotaxonomie der Pflanzen" Band 1-7, Birkhäuser Verlag, Basel, 1962-1986.

Krebs, H.Chr. "Recent developments in the field of marine natural products with emphasis on biologically active compounds" *Fortshritte Chem. org. Naturst.*, *49*, 151, 1986.

Lee, W.L., Ed. "Carotenoproteins in animal coloration" Dowden, Hutchinson & Ross, Inc., Stroudsburg, Pennsylvania, 1977.

Moore, R.E. "Volatile compounds from marine algae" *Acc. Chem. Res.*, *10*, 40, 1977.

Moore, T.C. "Biochemistry and physiology of plant hormones" Springer-Verlag, New York, N.Y., 1979.

Peter, M.G. "Chemical modifications of biopolymers by quinones and quinone methides" *Angew. Chem. Int. Ed. Engl.*, *28*, 555, 1989. A review on the structure and biogenesis of plant lignins, chitin, and sclerotized proteins of insects.

REFERENCES

Pietra, F. "Total synthesis of marine natural products: a powerful contribution to the understanding and development of marine organic chemistry" *Gazz. Chim. Ital.*, *115*, 443, 1985.

Scheuer, P.J., ed., "Marine natural products. Chemical and biological perspectives" Academic Press, New York Vol I-V, 1978-1983. General, multiauthor treatise.

Scheuer, P.J. ed., "Bioorganic Marine Chemistry" Springer-Verlag, New York, N.Y. Vol. I, 1987; Vol. II, 1988. General, multiauthor treatise.

Scheuer, P.J. "Marine toxins" *Acc. Chem. Res.*, *10*, 33, 1977.

GLOSSARY

absolute configuration-configuration that defines the chirality of the molecule, i.e. whether the molecule is like the left or the right hand. Old notations of absolute configuration are L- or D-, as related to L-glyceraldehyde or D-glyceraldehyde, to

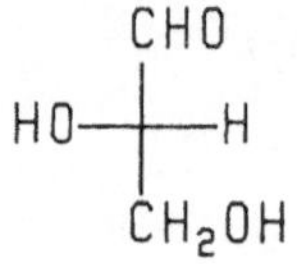

$$
\begin{array}{cc}
\text{CHO} & \text{CHO} \\
\text{HO}-\!\!\!\!-\text{H} & \text{H}-\!\!\!\!-\text{OH} \\
\text{CH}_2\text{OH} & \text{CH}_2\text{OH} \\
\text{L-glyceraldehyde} & \text{D-glyceraldehyde}
\end{array}
$$

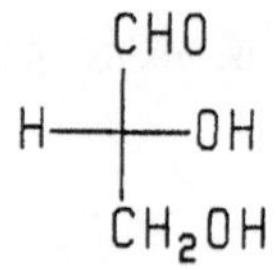

which the compound under examination has to be correlated by chemical transformations in order to assign the absolute configuration. This is shown above in Fisher projections, where the substituents at the vertical lines must be viewed below the plane and those at the horizontal lines above the plane of the paper. It is implied that the absolute configuration of glyceraldehyde is independently known via an absolute method. Recent notations of absolute configuration, R and S (from "rectus", right and "sinister", left) are based on arbitrarily assigned priorities to the substituents at the asymmetric carbon.

accessory pigments-pigments that harvest light and transfer the corresponding energy to the photochemical center in photosynthetic cells.

acetogenins-compounds that originate biogenetically from acetic acid (CH_3COOH) or from higher carboxylic acids derived from acetic acid.

acetylcholine activity-interference with the chemical transmission, due to acetylcholine, of impulses along nerve fibers of higher animals.

active hydrogen-hydrogen in a state more reactive than free molecular hydrogen (H_2), such as it occurs in photosynthesis and catalytic hydrogenations.

GLOSSARY

adrenalin (epinephrine)-hormone of the adrenal medulla and sympathetic nerves.

α-adrenoceptor blocking activity-inhibition of neurotransmitters by a drug. The synthetic drug chlorpromazine is quite active in this respect and is usually taken as a reference standard.

alcohols-an alcohol (R-OH) may be formally imagined to derive from a hydrocarbon (R-H) by substitution of a hydrogen atom with a hydroxyl (OH) residue.

alkaloids-nitrogen-containing metabolites that usually display basic reaction and originate biogenetically from amino acids. They include physiologically active compounds such as strychnine, morphine, nicotine, and caffeine.

alkyl group-residue (R) from a hydrocarbon (RH), resulting from the formal abstraction of a hydrogen atom.

amines-an amine (R-NR'R") may be formally imagined to derive from a hydrocarbon (R-H) by replacement of a hydrogen atom with the NR'R" residue, where R' and R' may be hydrogen atoms or alkyl or aryl groups.

amino acids-organic compounds in which the amino group (NR'R") and the carboxyl group (COOH) are in the same molecule. With α-amino acids the two functions are at the same carbon atom. Twenty α-amino acids in the L-configuration constitute the building blocks of the proteins. Natural products comprise many more types of amino acids, not necessarily of the α-type, or in the inverse absolute configuration with respect to protein amino acids.

anabolism-constructive metabolism.

anaerobic processes-biochemical processes that occur in the absence of molecular oxygen. Anaerobes are organisms that live without oxygen; certain organisms are facultative anaerobes.

analgesic drug-a drug that alleviates pain without blocking nerve impulses.

anion-an ion which bears a negative charge.

243

GLOSSARY

antiatherosclerotic drug-a drug that prevents the deposit of fats along the arteries.

aquaculture-exploitation of water media for the growth of seaweeds, invertebrates, and fish.

aragonite-a form of calcium carbonate ($CaCO_3$) which is harder, more dense, and thermodynamically less stable than the calcite form of common calcareous rocks. Aragonite is produced by some invertebrates (shells, pearls) and by certain calcareous algae. It is also formed as low-temperature surface deposits.

asystolic drug-a drug that induces a decreased contraction of the heart.

ataxia-an inability to coordinate voluntary muscular movements that is symptomatic of some nervous disorder.

atomic number-for the atom of a given element it represents the number of protons in the nucleus.

atomic weight-the weight in grams of an element relative to that of carbon taken as 12.

atoms-as far as chemistry and biology are concerned, the atoms are the ultimate building blocks of matter.

atriastimulant drug-a drug that affects the contraction of the upper part of the heart.

autotroph-a nutritionally self-sufficient organisms.

barophilic organisms-organisms growing at high pressures; generally they do not survive when brought to atmospheric pressure.

barotolerant organisms-organisms which grow best at atmospheric pressure, but survive under conditions of high pressure. In the case of unicellular organisms functions such as cellular division are inhibited at high pressure, however.

beam trawl-catching marine organisms on the sea bed by a tool made of a beam equipped with a bag-like net.

bile pigments-excretion pigments, originating from heme, which are eliminated from the liver and accumulated in the gall-bladder.

GLOSSARY

bioluminescence-emission of light of biological origin; in macroorganisms it is often due to luminescent bacteria.

biomass-total weight of living things in a given volume.

bioassay-determination of the relative strength of a substance (as a drug) by comparing its effect on a test organism (or organ) with that of a standard preparation.

biosynthesis-the synthesis of a natural product by a living organism or by a cellular culture, or by an enzymatic extract.

biotechnology-exploitation of living organisms, or their products, for technological purposes.

calcite-the softer form of calcium carbonate (see aragonite).

carbohydrates-compounds, also called sugars, of generally $C_n(H_2O)_n$ proportion of atoms which are produced in the photosynthetic process from CO_2 and H_2O. Comprised are mono-, oligo-, and polysaccharides, which are made up of one, a few, or many sugar molecules, respectively,

carotenoids-pigments, comprising eight isoprene-like units, and thus belonging to the tetraterpene class, which often occur in Nature as carotenoproteins, i.e. bound to a protein.

central nervous system-the brain and the spinal cord.

chemical bond-the bond that holds the nuclei together in a compound. The ionic bond results from the mutual attraction of ions of opposite electrical charges. The covalent bond results from electron sharing between the nuclei. The hydrogen bond results from a mixture of these two bonding modes. The Van der Waals bonding, which is the weakest bonding, results from the mutual attraction of instantaneous dipoles formed by instantaneous displacements of the electrons with respect to the nuclei.

chemical equilibrium-dynamic condition of a chemical system in which the rate of transformation of the reagents into the products equals that of the reverse reaction.

chemical mark-a compound characteristic of a specific homogeneous group of living or fossil organisms and which can be used for taxonomic purposes.

chemotaxonomy-a branch of biology that deals with the classification of living organisms on the basis of their specific chemical compounds.

chirality (*keir*, hand)-the property of a molecule of not having a superimposable mirror image.

chloroplast-a structure about 2.5 microns thick and 5 microns long, bound in a double-layered membrane, and composed of photosynthetic cells; it is the site of photosynthesis.

chromosome-thread-like part of the cell nucleus, made of DNA bound to a protein, which carries genetic information.

colonial organisms-organisms formed by aggregates of cells where neither organs nor tissue organization are discernible. Colonial should not be confused with multicellular; multicellular organisms have different cells for different functions.

color and structure-the carbon compounds that have a long series of alternate single and multiple chemical bonds absorb visible light and are thus colored. In contrast, when a molecule does not possess multiple bonds, or the multiple bonds are separated by more than one single bond, such as the polyunsaturated fatty acid EPA of diatoms, only high-energy, non-visible light is absorbed so that the visible spectrum emerges unaltered and the compound is colorless.

comparative biochemistry-a branch of biochemistry where similarities and differences in the biochemistry of the various taxa are emphasized.

concentration-the amount of a dissolved chemical compound per unitary volume. In order to make the measure independent of the nature of the compound, the concentration can be expressed in mole per liter.

configuration-for a molecule it specifies an arrangement of the atoms with respect to one another that can not be changed without breaking and reforming bonds.

GLOSSARY

conformation-for a molecule it specifies an arrangement of the atoms with respect to one another that can be changed by rotations around single chemical bonds.

coordination compounds-compounds made up of a central metallic atom linked to heteroatoms of molecules which are called ligands.

coral reef-a ridge formed by calcareous skeletons of living scleractinian corals and calcareous algae, and of their remains.

cross-linked polymers-in cross-linked polymers the polymeric chains are bridged to one another by chemical bonds; this reinforces the whole system.

de novo **synthesis**-synthesis of a natural product by a living organism starting from simple precursors.

diastereomers-stereoisomers which are not related to one another as mirror images.

dopamine-alkaloid involved in the transmission of nerve impulses in the brain; it is the biogenetic precursor of other neuroactive hormones (epinephrine and norepinephrine).

dredging-or dredge sampling- collecting samples of living organisms or sediments from the sea bottom by means of a dredge.

Earth's gases-methane and other simple natural gases.

electronegativity-it is a measure of the ability of an atom, or of a group of atoms, to attract electrons.

electron microscope-a microscope that uses a beam of electrons instead of a light beam to detect small objects. As the light associated with the electron has much shorter wavelength than visible light, the electron microscope has a higher resolution than the light microscope. The light associated with the electron, by impinging on a fluorescent screen, is made visible to the eye.

electrons-subatomic, negatively charged, small particles which travel around the nucleus at large distances (in terms of the dimensions of the electron and the nucleus) in such a number as to balance the positive charge of the nucleus.

GLOSSARY

embryology-branch of developmental biology that deals with the development of the embryo.

emulsion-mixture of two or more liquids where one of them occurs as non-miscible droplets of microscopic to submicroscopic size.

enantiomers-isomers related to one another as mirror images.

endoplasmic reticulum-system of membranes in the interior of the cytoplasm.

enzyme-a protein, or protein complex, that speeds up chemical reactions. Chiral products of enzymatic reactions generally are pure enantiomers.

epiphyte-a plant growing on another plant or on inanimate objects.

evolution-the process by which living things originate from earlier forms.

fatty acids-long-chain carboxylic acids of acetogeninic origin. They may be bound to glycerol to constitute the fats.

follicle cell-a small sac, chiefly a lymphatic gland.

fouling organisms-organisms that invade and obtrude other organisms or objects.

functional group-small, reactive portion of a molecule, the chemical behavior of which is largely independent of the type of molecule.

gel-liquid in which ultramicroscopic particles are dispersed or ordered. It may be elastic as gelatin or solid as silica gel.

genetic engineering-misleading term to indicate a branch of biochemistry that is concerned with the manipulation of genetic material. Actually, engineering is the application of scientific knowledge to problems of non-living systems.

gill nets-series of superimposed nets of different sizes that is suspended vertically in the water with meshes that allow the head of a fish to pass but entangle it as it tries to withdraw.

glycerol-the alcoholic portion of ordinary fats.

glycosides-acetals formed by the condensation (reaction with elimination of a water molecule) of the anomeric hydroxyl group (OH) of a sugar and the hydroxyl group of an alcohol.

248

GLOSSARY

hatching factor-a compound that affects the emergence of young from eggs or other structures.

hepatopancreas-glandular organ of invertebrates which is called liver in vertebrates.

heteroatoms-in coordination chemistry heteroatoms are atoms rich in electrons, e.g. oxygen or nitrogen.

heterotrophic bacteria-non-photosynthetic bacteria that can grow in specialized niches on various, even inorganic, compounds.

histamine effect-histamine is an alkaloid released under stress from tissues; it induces a dilatation of blood vessels, and thus a lowering of the blood pressure and inflammation of tissues.

hormones-compounds produced in tiny amounts in living bodies where they serve to regulate physiological functions.

hydrocarbons-compounds composed of carbon and hydrogen only. They are abundant in petroleum but are also biosynthesized by living organisms.

hydrolysis-the splitting of a compound, such as a polysaccharide, by reaction with water to give smaller units (in this case the component mono- or oligosaccharides).

indomethacin-synthetic indole alkaloid; it is taken as a reference standard for the analgesic and antiinflammatory action of other compounds.

invertebrates-animals that lack a vertebral column.

ion-an atom which has lost or gained one or more electrons and which therefore carries positive or negative charges.

ionophoric antibiotics-macrolides that have the ability to sequester an ion, or a complex ion, by placing it at the center of their macrocycle, coordinated to heteroatoms, generally oxygen atoms.

irregular terpenes-terpenes that can not be formally dissected into isoprene-like units; they do not follow the isoprene rule.

249

GLOSSARY

isoprene (2-methyl-1,4-butadiene)-for simplicity it may be considered as the formal precursor of the terpenes; the true precursor is mevalonic acid, however.

isoprene rule-the regular joining together of intact isoprene-like units to give compounds made up of two to eight of such blocks.

isotopes-atoms having the same atomic number and thus constituting the same element while differing in the mass (i.e. in the number of neutrons in the nucleus).

K⁺,Na⁺-ATPase-enzyme active in the transport of potassium and sodium ions across cellular membranes.

Le Chatelier principle-a system reacts to an external stimulus by moving in the direction in which the effect of the applied stimulus decreases.

lipids-a class of diverse compounds, of animal and non-animal origin, which include fats.

macrolides-compounds having a large lactone (cyclic ester) ring.

magnesium-one of the alkaline-earth metals; it is the central metal in chlorophylls and bacteriochlorophylls.

metabolism-it is the dynamic condition of the compounds contained in organisms. It includes all enzymatic processes that occur in the cell to allow growth and reproduction.

miclosclere-minute sclerite, i.e. sclerotized plate, in sponges.

mixed biogenesis metabolites-natural products deriving from two, or more, different biogenetic routes.

mole-it is the amount of a compound numerically equal to its molecular weight.

molecular label-an isotope that can be distinguished from the other isotopes of the same element, usually through its radioactive or magnetic properties. Via synthetic or biosynthetic technologies, atoms of a molecule may be replaced by such a molecular label.

molecule-two or more atoms bound together by chemical bonds to constitute the smallest units into which a pure compound can be separated.

morphine-opium alkaloid with potent analgesic activity but with deleterious side effects. It can be chemically transformed into heroin, which has a much higher euphoric effect.

net production-the part of the primary production that is left after losses from respiration.

neurotransmitter-a compound that diffuses across the synaptic cleft through nerve cells.

nitrogen cycle-other than through nitrogen fixation, biological transformations of nitrogen compounds involve oxidation of soil ammonia by nitrifying bacteria, first into nitrites (NO_2^-) and then into nitrates (NO_3^-). Soil ammonia is released in the decomposition of organic material.

nitrogen fixation-process in which nitrogen-fixing bacteria and certain cyanobacteria transform molecular nitrogen (N_2) into nitrates (NO_3^-).

nucleosides-compounds resulting from the joining of a sugar molecule with a molecule of a nitrogen-containing aromatic base. By phosphorylation they yield nucleotides. Various types of free nucleosides are found as secondary metabolites.

nucleotides-nucleoside phosphates, four types of which are found in the nucleic acids.

nucleus (of the atom)-central small part of the atom with a positive electrical charge and practically the whole of the mass of the atom.

nucleus (of the cell)-in eukaryotic cells it is a specialized structure enclosed in a double-layered membrane; it contains genetic material.

one-carbon biomethylation-methylation of electron-rich atoms in molecules of living organisms by S-adenosyl-L-methionine.

optical activity-plane-polarized light is rotated by the solution of a chiral compound present either as a single enantiomer or in enantiomeric excess. The instrument which measures the extent of rotation is called polarimeter. Looking toward the light source, through the solution, clockwise or counterclockwise rotations are indicated with the plus or minus sign, respectively. The extent of the rotation is proportional to the amount of enantiomer in the sample tube and depends on the wavelength of the light. No prediction of absolute configuration can generally be made from the sign of the optical rotation. A more complex instrument, the dichrograph, uses circularly polarized light which, as a chiral light, recognizes opposite enantiomers. This technique is called circular dichroism.

organelles-specialized substructures of cells.

oxidation-oxidation of a chemical compound involves removing electrons from it. With organic compounds, oxidation often results in removing hydrogen atoms, such as it occurs in oxidizing an alcohol (R_2CHOH) to a ketone (R_2CO). Reduction is the reverse process.

oxygenic photosynthesis-conversion of water (H_2O) and carbon dioxide (CO_2) into carbohydrates ($C_n(H_2O)_n$) and other essential metabolites under the action of light with production of molecular oxygen (O_2).

parasite-a living organism that is associated with another living organism and benefits from it, generally not attempting to kill it.

penicillins-antibacterial alkaloids produced by molds of the genus *Penicillium*.

peptides-compounds formed by the condensation of a few amino acids (see peptidic bond).

peptidic bond-bond formed by the condensation (reaction with elimination of a water molecule) of the amino group (NHR) of an amino acid with the carboxylic group (COOH) of a second amino acid.

periodic table of the elements-arrangement of the elements in horizontal periods and vertical groups according to increasing atomic number. The elements of a group have similar chemical properties whereas from the left to the right along a period there is a change from metallic to non-metallic properties.

Petri dish-round, flat, glass container used for the culture of microorganisms.

phenols-derivatives of benzene (C_6H_6) with a hydroxyl group (OH) replacing a hydrogen atom.

GLOSSARY

photochemical reactions-chemical transformations that begin with absorption of light by a molecule, which is thus raised to an electronically excited state. This may be followed by deactivation to the initial state with light emission, or by a chemical transformation in a dark process, or even by transfer of energy in a collision with a second molecule which is thus raised to an electronically excited state. In the latter case, the first excited molecule acts as a sensitizer.

phylogenetic mark-chemical compound characteristic of organisms which belong to phylogenetically related groups.

phylogeny-evolutionary history, or evolution, of a homogeneous group of organisms.

phytoalexins-compounds produced by plants under physical or biological stress and which act as antimicrobials.

plankton-the floating or weakly swimming bacterial and minute animal and plant life of a body of water.

plasmids-extrachromosomal genetic material of bacteria.

plastids-membrane-enclosed organelles of photosynthetic cells; those containing chlorophyll are called chloroplasts.

platelet-aggregation inhibitory action-ability of certain natural products to hinder the coagulation of blood by preventing platelet aggregation.

polar compounds-compounds which have polar functional groups (i.e. groups formed with atoms of largely different electronegativity). Polar compounds are soluble in polar solvents. Non-polar compounds, typically the hydrocarbons, lack polar groups and are soluble in non-polar solvents, typically other hydrocarbons. A polar compound is strongly retained by polar chromatographic substrates, such as silica gel or alumina, and requires solvents with hydroxyl functions, or other polar functions, to be eluted.

polymers-compounds of high molecular weight, consisting of repeating structural units that result from the union of simpler monomers.

253

polypeptides-compounds of the type of the peptides, but formed from a higher number of amino acids.

polyphenols-highly oxidizable phenyl (C_6H_5-) or polyphenyl compounds with two or more hydroxyl groups (OH) replacing hydrogen atoms at the phenyl rings.

polysaccharides-see carbohydrates.

pressure-force per unitary area.

primary metabolism-biochemical reactions that start from carbon dioxide and light to afford primary metabolites (sugars, amino acids, and nucleotides).

primary production-the production of compounds through photosynthesis.

principle of maximum separation of electron couples-owing to electron repulsion, the electron couples of chemical bonds and lone pairs on atoms tend to lie as far apart as possible from one another in molecules. This determines the geometry of the molecule.

prosthetic group-the non-protein group in a conjugated protein.

proteins-compounds of the type of the polypeptides but consisting of a higher number of amino acids, which are chosen from only twenty types of α-amino acids. Conjugated proteins are made up of a protein and a loosely bound prosthetic group; on mild treatment, the free protein and prosthetic group can be obtained.

pyrrolic pigments-α-linked polypyrrole pigments typical of bacteria.

racemate-a fifty to fifty mixture of the two enantiomers of a chiral compound.

receptors-specialized cells which have the ability to convert a stimulus into a nerve impulse. Receptors and their accessory structures constitute sense organs. Sense organs may react to various external stimuli, such as sound, light, or chemical compounds, usually with high specificity resulting from the combination of various hand-glove systems or a more elaborated strategy.

reduction-it is the reverse process of oxidation; see oxidation.

saponins-compounds composed of a steroid or triterpene bound to a sugar by a glycosidic bond.

GLOSSARY

secondary metabolism-biochemical reactions that start from primary metabolites to give natural products (= secondary metabolites) which are specific of certain groups of organisms or even of a single species.

siderophore-a compound, generally of microbial origin, which has the ability to transport iron.

sodium channels-to import nutrients, the cell has to export some substances in order to maintain a material balance. The cell membranes act as selective filters and in normal animal cells the imported substances are potassium ion, glucose, and amino acids, and the exported substance is sodium ion. When dinoflagellate toxins, such as saxitoxin and gonyautoxins, are ingested, they dissolve into body fluids as cations, thus being electrostatically attracted by negatively charged groups located around the orifice of the sodium ion channel. The channel is thus blocked, which prevents the cell from importing nutrients and causes its rapid death.

stereoisomers-compounds differing in the spatial positions of the constituent atoms, which are bonded to one another in the same order.

steroids-irregular triterpenoids which generally have lost some carbon units; certain sponge sterols have also acquired some carbon units.

structural isomers-two or more compounds that are made up of the same number of atoms of the same elements but differ in the sequence the atoms are connected to one another.

surfactants (= **detergents**)-compounds that affect the surface tension of a liquid.

symbiont-a living organism that is associated for mutual cooperation with another living organism.

taxon-unit in biological classification. Taxa are arranged in ascending order from species to superkingdom.

terpenoids-compounds that can be formally imagined to derive by the joining together of five-carbon isoprene-like units. The number of such units indicates the class of the terpene; thus, hemiterpenes, monoterpenes, sesquiterpenes, diterpenes,

GLOSSARY

sesterterpenes, triterpenes, and tetraterpenes are formed from two, three, four, five, six, and eight isoprene-like units, respectively.

tetrahedral, trigonal, and digonal carbon-bonding modes of carbon to either four, three, or two partner atoms which lie at the vertices of a tetrahedron, the vertices of a triangle, or along a line, with carbon at the center.

total synthesis-laboratory synthesis of a natural product starting from simple precursors.

truncated (or degraded) terpenes-terpenes that have lost at least one carbon atom during biogenesis.

X-ray diffraction-light being reflected from a ruled surface, such as a grating, may produce fringes of parallel light of high luminosity and low luminosity, while decomposing into its component colors. This is the phenomenon of diffraction of light. With light of very short wavelength, such as the X-rays, diffraction can only occur with such tightly ruled surfaces as those created by the network in crystalline matter. In this case, diffraction bands allow us to measure the interatomic distances and to reconstruct the relative positions of all atoms in the crystal.

yeasts-unicellular phase of many lower fungi where reproduction occurs by fission or budding.

INDEX

The location of illustrations is indicated by page numbers in boldface.
The location of structural formulae is indicated by page numbers in italics.

INDEX (SEAWEEDS-SEAWEED PRODUCTS)